GW00419359

PERMANENTLY WITHDRAWN

INTRODUCING
Benchwork

INTRODUCING
Benchwork

STAN BRAY

Patrick Stephens, Wellingborough

First published in 1986

British Library Cataloguing in Publication Data

Bray, Stan
Introducing benchwork—(PSL model engineering guides; 3)
1. Engineering models
I. Title
620′.00228 TA177

ISBN 0-85059-742-0

Patrick Stephens Limited is part of the Thorsons Publishing Group

Photoset in 10 on 11 pt Univers by MJL Typesetting Services Limited, Hitchin, Herts. Printed in Great Britain on 100 gsm Fineblade coated cartridge, and bound, by Butler & Tanner Limited, Frome, Somerset, for the publishers, Patrick Stephens Limited, Denington Estate, Wellingborough, Northants, NN8 2QD, England.

CONTENTS

INTRODUCING BENCHWORK

Model engineering as a hobby is a mixture of skills. These include various types of machining and also quite a large amount of hand work. The hand work is diverse, including the ability to cut, smooth, and join metal. These processes can be executed in a variety of ways and it is the object of this book to help readers to carry out the necessary operations.

Before any work can be done it is necessary to have an idea to work to. It is possible to get this from your own head and many good models are made in this way, the design being stored in what must be the world's finest computer, the human brain. Sometimes though you will wish to copy a published design, or to work from an original design drawn by the manufacturer of the object being modelled. This will involve reading those drawings and translating them into the third dimension. The drawings must be transferred into some form of markings on the material in which you are modelling and you will find the necessary techniques explained in the chapters on reading drawings and marking out.

The next stage will be to cut and shape the components of the model. The parts will then have to be assembled which may be done by a variety of mechanical means or by means of heat. Read the instructions for these processes carefully before carrying out the work.

Some advice on the types of material likely to be used in modelling is included and tables give guidance for selecting the correct material sizes and making the correct size holes for any threads that might need to be made.

The all important thing in making a model is to care about the finished product. The parts must fit properly together for a model to work and the finish must be right—care in finishing will repay itself one hundred-fold. Most modellers have at some time fallen into the trap of hurrying to finish a model, only to be very disappointed with a result which might have been so much better. Hopefully this book will teach you the skills required to be a successful modeller, what it cannot do is to give you the patience to finish the job properly—you must have the will power to provide that yourself. If you feel that a model must be finished for Christmas, then make it next Christmas. That way you will be happy with it for many Christmasses to come. Hurry it and it will end up in a cupboard out of sight, rather than displayed proudly where all can see it. Above all make sure that you only work on the model when you are in the mood, or working on it will become a chore. Follow this advice and success will be yours.

1 *THE BENCH*

The bench on which you will work is, needless to say, of the utmost importance. A decent quality bench will repay itself many times over, if it is not up to standard in no time at all it will need repair or replacement. There are quite a few manufacturers who supply benches and if one of these would suit your purpose then it is worth consideration—the majority are designed with the woodworker in mind, but some suitable for metal work are available. Many people choose to make their own bench so that it can be specially designed to fit the space available.

Careful thought should be given to the siting of a bench. It should, if possible, be bolted to a wall and all edges where it meets walls should be sealed. It is quite amazing how nuts, bolts, taps, drills and other small items will find their way down the back if this is not done. A simple wooden batten fixed with a few pins will suffice and save hours of frustration later. If possible site the bench near the light and have at least one electric power point nearby.

If you plan to make your bench it is important to use timber of adequate strength for the legs and cross braces. If it is available an alternative is to use slotted angle or any form of angle iron. There should be some shelf space available underneath, but it might be worth considering leaving room for your legs to go under so that you can sit down. The bench legs and braces should be made by bolting together with coach bolts rather than using wood screws which, after a period, work their way loose and cannot be tightened up again. A coach bolt fitted with a washer under the nut is less likely to loosen and if it does it can easily be retightened.

Right—*Method of constructing a bench with wooden framing, leaving the maximum space underneath for storage purposes.*

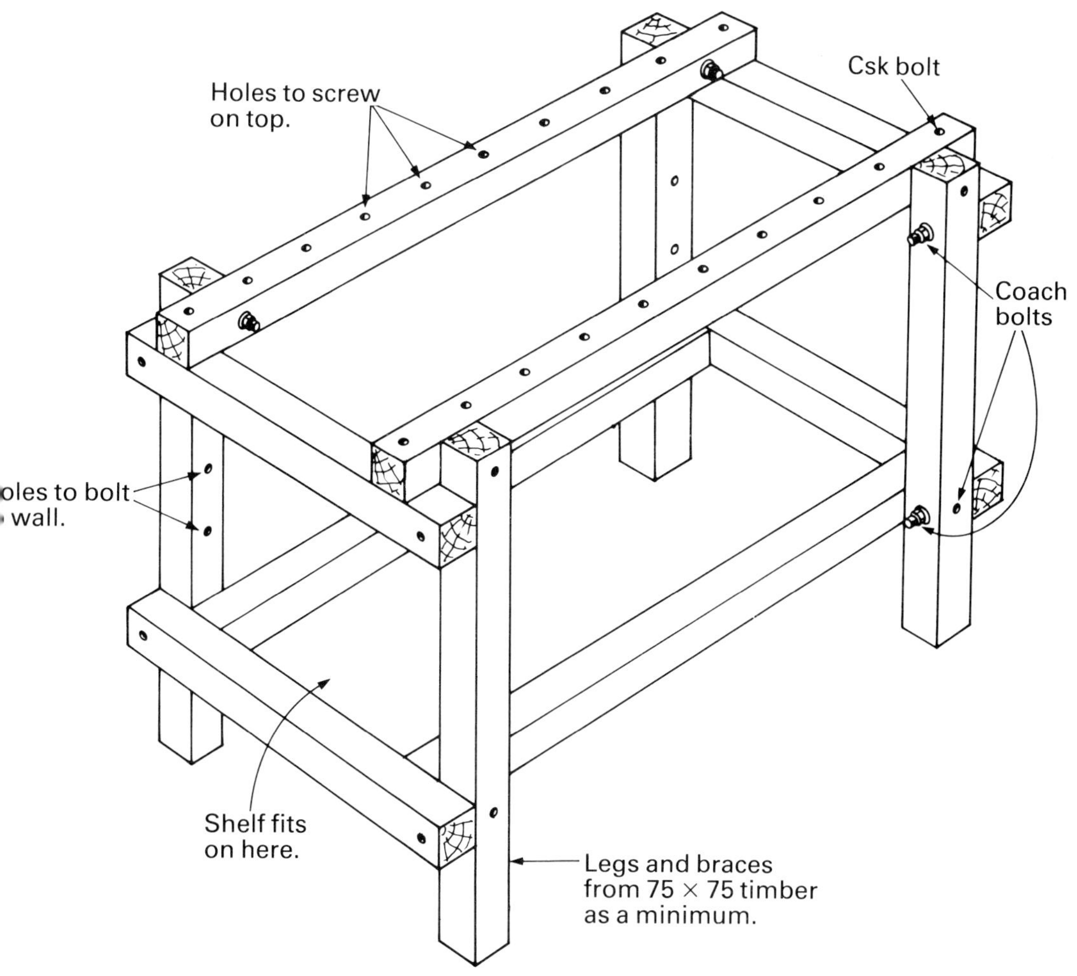
Holes to screw
on top.
Csk bolt
Coach
bolts
oles to bolt
wall.
Shelf fits
on here.
Legs and braces
from 75 × 75 timber
as a minimum.

A typical bench in a home workshop

The top of your bench must be made of stout timber, preferably a single piece as using planking leaves cracks. If planking has to be used then it is advisable to lay chipboard over the top to give a suitable surface. If the top is of wood it will need a coat of varnish to prevent any oil that is spilled from soaking in. At one time it was customary to cover benches with linoleum but this seems to have gone out of fashion. Whatever your personal preference it is vital that the surface can be easily cleaned.

In my opinion a bench should be between 18 in and 2 ft wide—if it is wider than this it tends to become rather cluttered at the back. The ideal height will depend on how tall you are. To have the top of the bench at hip level so that the top of the vice is at waist height is usually found comfortable. You will find that you require as much storage space as possible within easy reach of your bench so that you do not have a route march to get at tools.

Right—*Coach bolts such as these should be used to hold a bench framework together in order to give it strength.*

Below—*Where coach bolts cannot be used, such as for reasons of room, then coach screws provide the next best choice. They have the advantage over ordinary wood screws of being able to be tightened up more, due to the shape of their heads, thus leaving less room for movement of the framework.*

Opposite—*A typical engineer's vice. This one has a swivelling base which makes it more versatile. Note the soft jaws, to protect work from damage.*

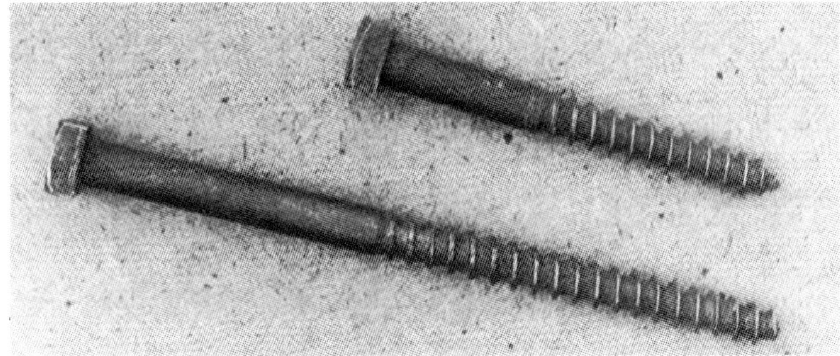

A vice is a necessity for model engineering and an engineer's type is the most suitable. These can be bought in a variety of sizes and you would be well advised to procure the biggest that you can afford, or have room for. Some models have a swivelling base which can be very useful. The vice should be bolted to the bench, rather than held down with wood screws. Vices are made of cast iron and should not be used to hammer on as, no matter how big and strong they seem at first glance, they are liable to crack. A metal block of some sort should be available for this purpose. An anvil is ideal, but space for such a luxury may well be a problem. I use a block of steel which was obtained from a scrap yard, and an old piece of railway line for very heavy work. The latter, whilst somewhat bulky, is very strong and will withstand any amount of ill treatment. For work which is too tiny to be held in the vice, some sort of auxiliary arrangement is needed—ideas for making various small devices suitable for this purpose will be found later in the book.

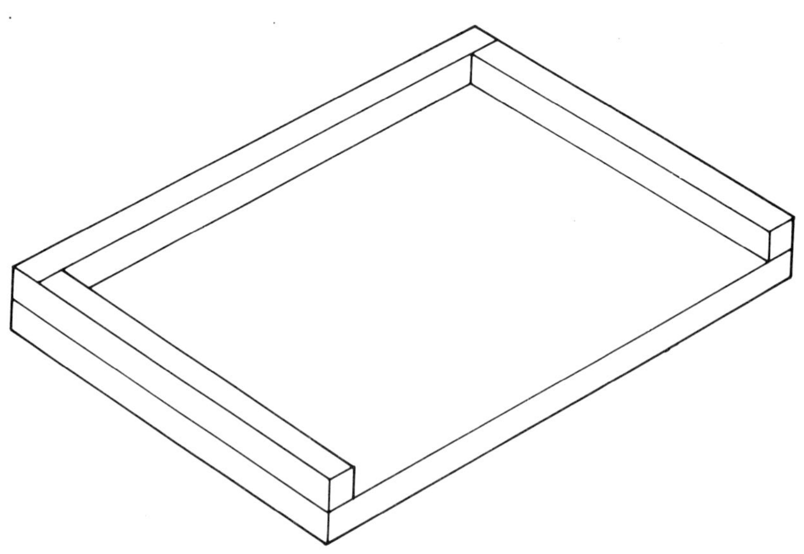

If your work is to be of a very fine nature then a separate tray with raised edges to clamp on the bench is useful to prevent small parts from going missing—it will save a lot of frustration! It is a good idea to cover it with green baize, this can be bought with a self adhesive backing, and is an ideal material for fine work. It can be cleaned quite easily with a quick brush down.

If you cannot have a permanent bench then something portable will have to be designed. Vices can be purchased that will clamp into position on a table. (They can also be useful as an extra vice where a permanent one is in position, and for small work can be raised on wooden blocks to a suitable height as fine work often needs to be done at a higher level than does work of a heavier nature.

Above all make sure that the area where you are going to work is as comfortable as possible. Comfort will mean that you will be able to work much more accurately. It will also mean that the hobby will remain a pleasure, rather than turning into a chore. Make sure that things are kept clean and tidy. Clearing up after each session pays dividends in the long run usually enabling you to work faster and with more pleasure.

Left—*When hammering, do not lay the object on which the hammering is to take place directly on the vice as this could cause considerable damage. Instead keep a block of steel that can either be laid on the bench, or secured between the vice jaws, and work on that.*

Below left—*A simple wooden tray with three sides, possibly covered with green baize, will help when doing very fine work.*

2 *WORKSHOP DRAWINGS*

It is almost certain that in order to get the best from the hobby of model engineering a knowledge of reading workshop drawings will be required. Drawings are made in order to reduce an idea for a model to a flat plan, in other words, to make what will be a three dimensional article understandable in two dimensional form. Drawings, or plans as they are sometimes called, appear regularly in magazines and there are a large number of firms that will supply them. It is quite possible to get a drawing for virtually everything you might want to make. Such drawings might not be to the scale that you require, but that takes us into another area, and so for the purposes of this chapter I must assume that the drawing you require is available in the scale you want. This does not mean that the actual drawing will be the exact size you require but that the measurements shown on it will be right for you. Sometimes it is not possible to produce a drawing as large as the object being made and at other times to produce the drawing to the exact size would make it too small to read. For example, drawings for model cylinders are frequently produced twice full size in order to show the details clearly. When drawings are reproduced the paper may stretch or shrink unpredictably, so nobody can be quite sure what size the actual drawing will be. Because of this measurements will be shown on the plan and you must work to these rather than try measuring the actual drawing. If any measurement is not shown then it should be possible to calculate it quite easily from other measurements.

Workshop drawings for industry conform to a laid down standard. An engineer should therefore be able to pick one up and read it without any problem. When it comes to drawings for models, things get a little more difficult. Many model drawings are produced by people who have not had proper training in engineering drawing—they usually have some knowledge of the subject, but a little knowledge can be a dangerous thing. I have on several occasions had skilled engineers who read drawings all day long ask me how to interpret a something on the drawing for a model. Not that I know more about reading drawings than they do, far from it, they could lose me in an engineering workshop, but model drawings are something different. The main problem lies in the fact that model drawings often employ a mixture of drawing techniques. These techniques are explained below.

The first type of drawing you might come across is known as an

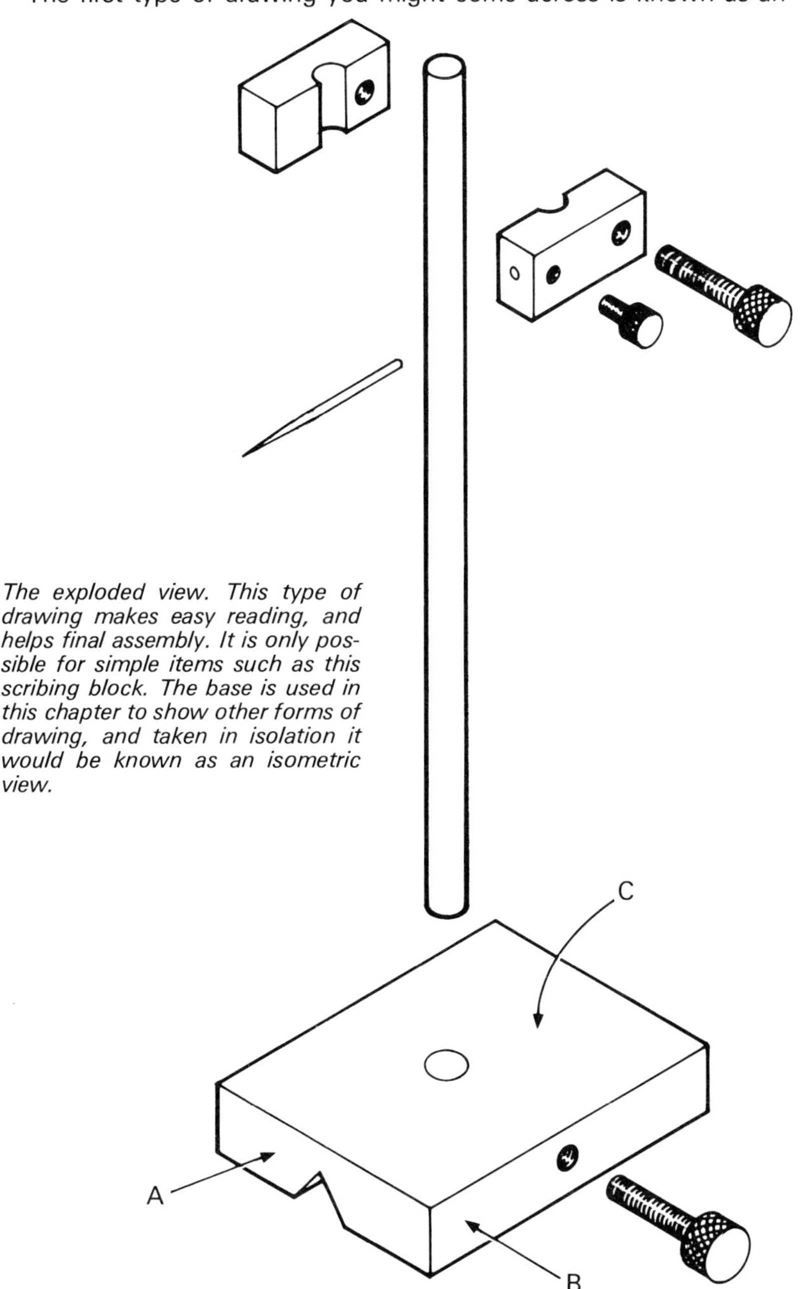

The exploded view. This type of drawing makes easy reading, and helps final assembly. It is only possible for simple items such as this scribing block. The base is used in this chapter to show other forms of drawing, and taken in isolation it would be known as an isometric view.

exploded view. Such a drawing is more likely to be seen in a magazine than bought as a plan, and although they are fairly rare in model engineering, they do happen. Model aircraft builders use this type of drawing all the time and with great success. They are self explanatory, but are only suitable for very simple components. They have the advantage that you can see how to assemble things, and in my opinion are ideal for a beginner. The one shown here is of a simple scribing block. Scribing blocks will be dealt with in another chapter, but they are an easy subject for the beginner, and none could be easier than the one shown. It can be seen that the gadget consists of a metal base with a vee in the bottom. A hole in the centre takes a metal rod, and a screw is provided to lock the rod in position. At the top are two metal blocks with half round cut-outs to enable them to be clamped to the metal rod by means of the screw which is shown. The only other part is a small scriber, which again is held with a screw. No measurements have been shown as the design is not at this stage intended as a project, but normally measurements would be shown alongside each component.

As you can see, that type of drawing is very easy to understand. Imagine though that there were twenty or thirty different parts, these could not possibly be shown in that way and so a different method is employed. The draughtsman produces what is known as a general arrangement, which shows the whole thing. Each part is then drawn separately, everything being assembled in accordance with the general arrangement, which is invariably known as the 'GA'. By labelling the parts both on the 'GA' and when they are drawn individually, things are fairly easily organised. The problem is how to show each component, which is why a standard was introduced. In order to explain this I shall use the base of the scribing block as an example.

The second method of drawing is known as isometric projection, and consists of showing all horizontal planes at an angle of thirty degrees, whilst the vertical ones remain vertical. In fact if we look at the base in the exploded view, it has been drawn isometrically. The result is as near a three dimensional picture as we can get. It has the effect though of making holes look oval, and distorting some lines, and one cannot see what is underneath. Suppose for example that at the bottom was another hole to allow the base to fit on to something else. There would be no way of showing it. Once again then we have a drawing which is ideal for simple things but if the subject is just the slightest bit more complicated it will not do. Nevertheless you will find this projection used in model engineering and you will find it nice and easy to work to.

The third method is to take the component, (again the base of the scribing block) and to show it in sections. This is called orthographic projection and is the way that all engineering drawing is carried out. There are two methods of doing this. The English way is known as first angle projection. If we look at the drawing on the top left we see the base from one position. All we would actually see would be a plain metal block with a vee in the bottom. Neither of the holes would be visible. We see these in this case, as all hidden detail is shown in dotted lines. The

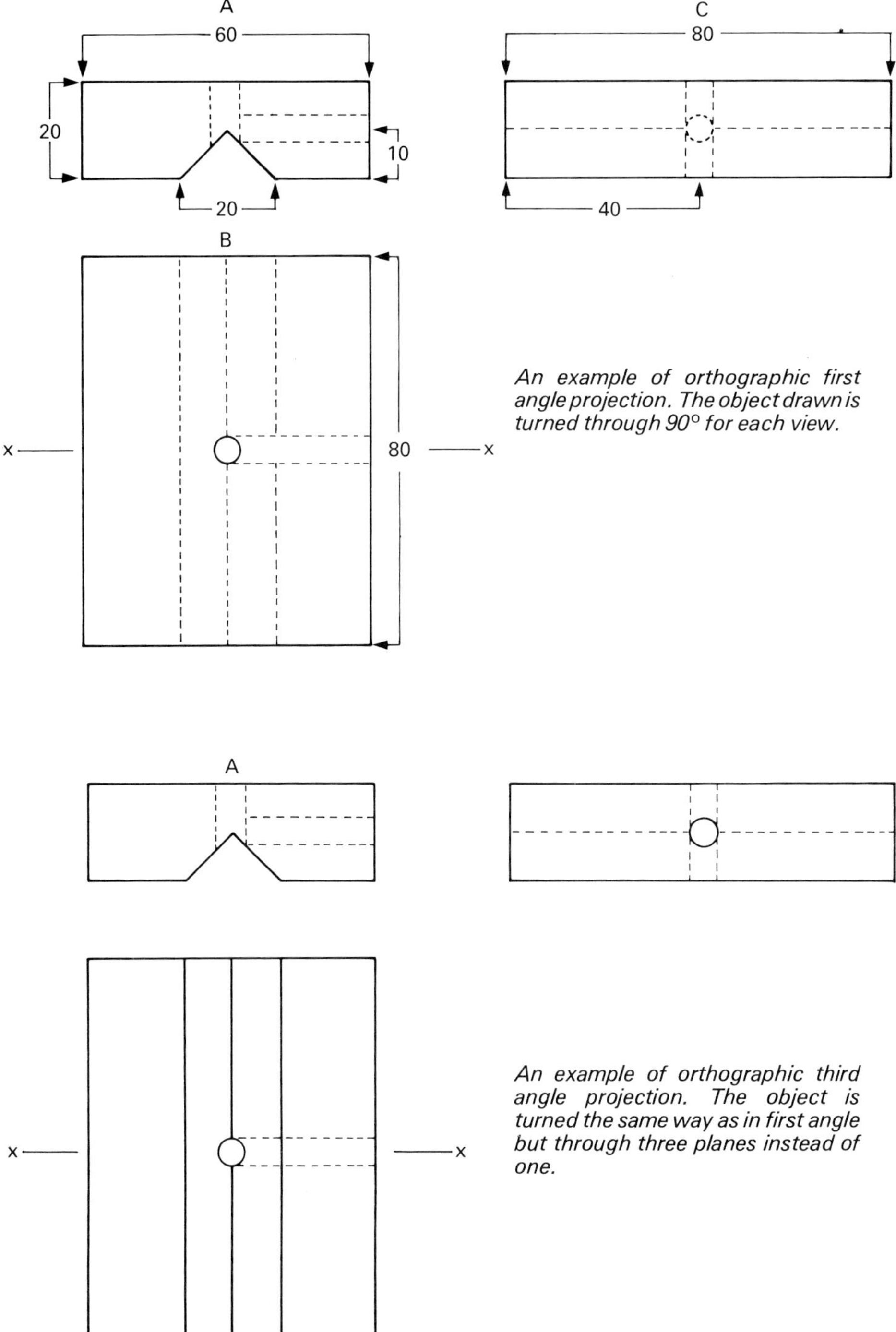

An example of orthographic first angle projection. The object drawn is turned through 90° for each view.

An example of orthographic third angle projection. The object is turned the same way as in first angle but through three planes instead of one.

view to the right of that shows the component turned round once to the right. We cannot see the vee, we cannot see the holes. All these are shown by dotted lines, but now the hole in the side becomes a different shape because of the different angle. Underneath the first view, the block is shown as turned to the top. We see the centre hole clearly. The vee and the other hole are dotted. If the block was complicated then a fourth view could be shown to the left of the first view but in this case we have enough detail with the three views to enable us to make it. Each view shows the block turned once, hence the name first angle projection. You will see that for the first angle view, dimensions have been shown, the convention being that the dimension lines are thinner than the outlines, as shown here.

Orthographic projection can also be made in what is termed third angle. This is basically the American way, but you will most certainly find it used in model engineering. It works exactly the same as first angle, except that the component is turned through three planes each time. This means that the opposite side is shown. Comparison of the

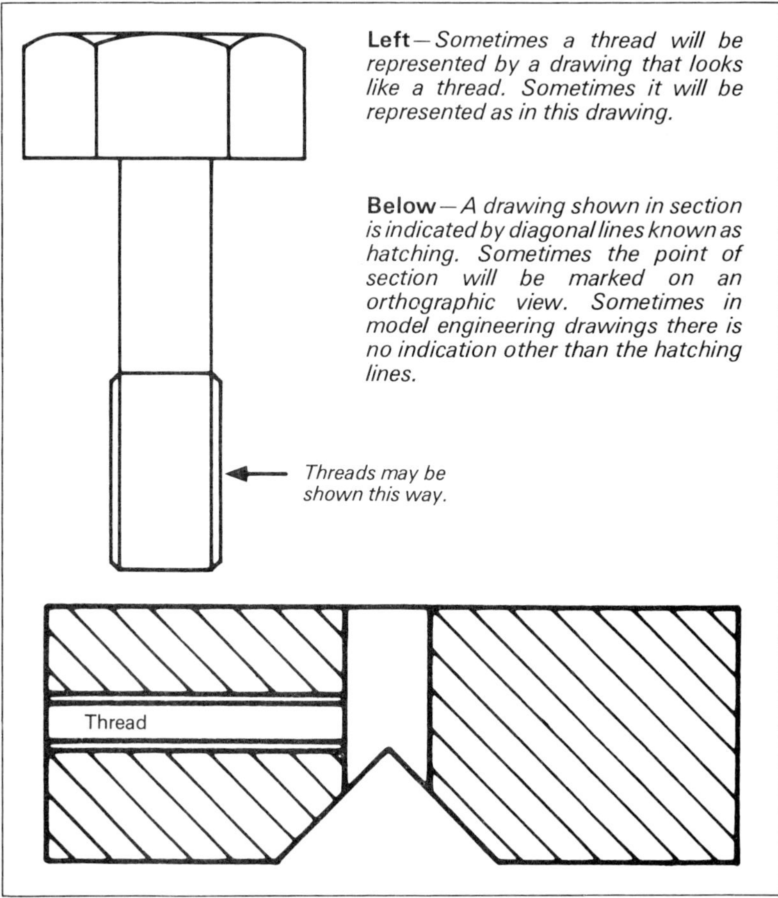
Threads may be shown this way.
Thread

Left—*Sometimes a thread will be represented by a drawing that looks like a thread. Sometimes it will be represented as in this drawing.*

Below—*A drawing shown in section is indicated by diagonal lines known as hatching. Sometimes the point of section will be marked on an orthographic view. Sometimes in model engineering drawings there is no indication other than the hatching lines.*

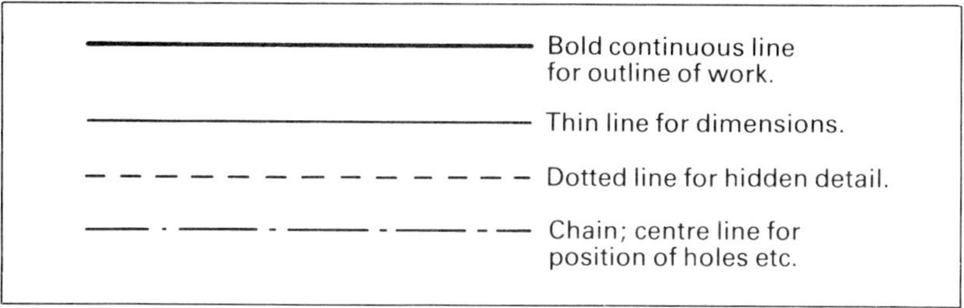

This shows the type of lines seen on workshop drawings and what they indicate.

drawings is the best way to see what is meant. The result as far as we are concerned will be the same, except that the view underneath the first one will be the base, and the one on the right will be the opposite to the one in a first angle projection.

Sometimes you need to see what a component would look like if cut in half. This is called sectioning, and when a component is shown in section the fact is shown by diagonal shading. The lines can go either way, and quite often you will find lines going in opposite directions, in order to separate areas of the drawing. It may well be that a line will be shown on an orthographic projection to indicate the point at which the section is taken. It will be marked with an 'x' at either end. The people who do model engineering drawings seem to be rather fond of sectioning, and a great many drawings are done that way when there is no need which can be a little confusing at times.

Now you have got some idea of how the basic drawings will look there are one or two other little points that you might need to know. If there are rows of holes then frequently these will be joined up with a line consisting of long dashes alternating with dots. This is known as a centre line and is only there as a guide, to show that all the holes should be on the same line. Small details of a component will often be shown as an additional drawing. Threads are shown sometimes in a manner that looks like a thread and at other times as two lines. The sketch illustrates what I mean.

Basically that should cover all that you need to know to make a model. Beware though as the drawings supplied often are not always straightforward. Some designs mix up their projections while some show just one face of a projection in some places and two or three in others. Study the drawings very carefully in conjunction with what I have written and you should not have too many problems.

3 *MARKING-OUT TOOLS*

Marking out basically means transferring the details from the drawing to the metal with which you are going to work. This serves two purposes—lines to indicate where to cut or machine and marks to show where to put the necessary holes. Lines are made with a scriber and marks for holes with a centre punch. It is essential to be accurate when marking out so considerable care needs to be taken in measuring. Apart from a scriber and centre punch there are quite a few other aids available and I shall describe these first of all, before going on to explain the process of marking out.

THE SCRIBER

The scriber is a long, thin, pointed piece of hardened and tempered steel. The handle is probably knurled to enable it to be gripped. They are made in a variety of sizes and shapes, although the pointed part should be the same whatever the shape of the main body of the instrument. It is possible to purchase them with a hooked end, and these are known as engineer's scribers, as the hooked part allows extra pressure to be put on the scribed line. Some have retractable points to allow them to be put in the pocket, and another type has a flat blade at one end, which is mainly

Three types of scribers. The top one is the normal type purchased in engineering suppliers. The second has a retractable point, and a clip so that it can be kept in the breast pocket. The bottom one has a flat end and, whilst mainly used by woodworkers, can be useful to the model engineer.

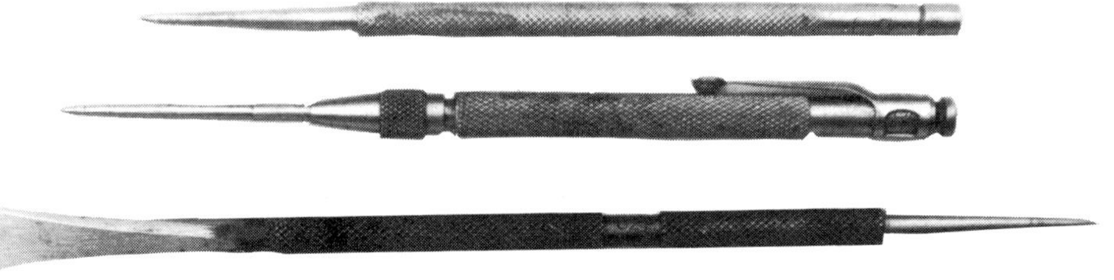

used for work on wood. It is possible to make them at home quite easily. The point on a scriber is at 30°. They can be sharpened when blunt, either by very light pressure on a fine grindstone, or preferably by rotating on a well oiled oil stone. They do need to be kept very sharp if they are to function properly. They should not be used as bradawls, centre punches, or anything else, but should be kept for the purpose for which they were designed.

CENTRE PUNCHES

Centre punches come in two types, or at least in theory they do, but these days it is getting increasingly difficult to purchase one type. The punches consist of a short length of tool steel with a parallel body, tapering near one end, and then the taper sharply increasing into a point at the end. The point on a centre punch should be at 90° as this enables a drill to set into it without wandering. Another type of punch, frequently called a centre punch but for which the correct name is a dot punch, has the point at 60°. The finer point enables it to be used more easily for locating the point on the metal where the hole is to go, therefore a dot punch should be employed first and a centre punch then used to increase the angle of the mark in the metal for easier location of the drill. In point of fact the centre punch is now the one which is difficult to purchase, and this seems somewhat strange. At one time dot punches

Below—*The drawings show the differing angles of the points of dot and centre punches.*

Bottom—*An automatic centre punch. It is adjustable in strength, and works when the spring loaded top is pushed in.*

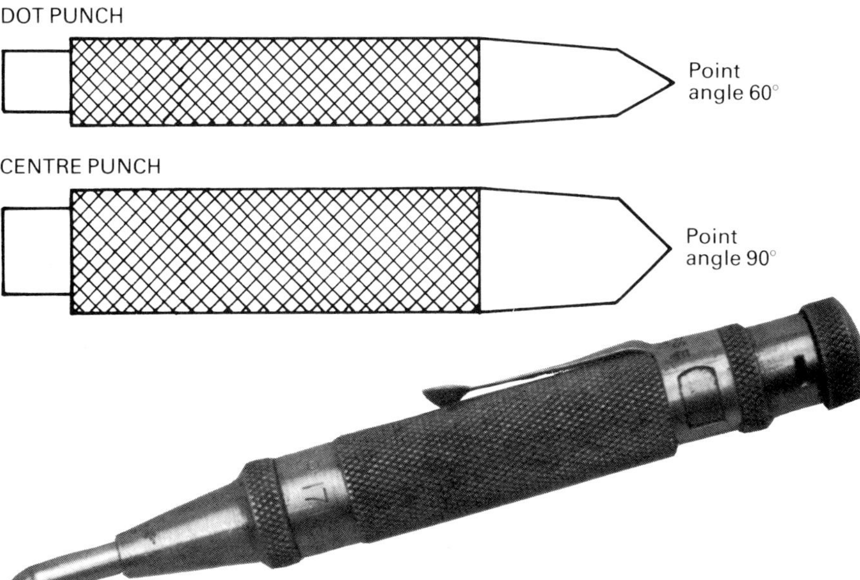

were sold only in small sizes, nowadays punches with points at 60° come in quite large sizes, all sold under the name of centre punch. The body of the punch might be round, in which case it should be knurled, or it might be square or hexagonal in section. It is possible to buy punches with 60° points that work automatically, in other words instead of hitting them with a hammer you just press them. They are spring loaded inside and this causes them to strike the metal with quite a heavy blow. The strength of the blow is usually adjustable. They have their uses in certain situations but remember that the mark they make should be opened out with a proper centre punch afterwards.

RULERS OR RULES

My maths teacher used to insist that a ruler governed a country, and a rule measured, but even so the objects I am talking about are more often than not sold as rulers. I suppose that everybody is familiar with a ruler of some sort or another, but the type required for model engineering needs to be made of steel. The gradations on it need to be marked with very fine lines, if the lines are thick and heavy you could easily end up with a comparatively large error by measuring from the wrong edge of the mark. The rule should be graduated with very fine divisions, say 1/64 of an inch if using imperial measurement, or with 0.5 of a millimetre if metric. This does not necessarily mean that the whole ruler needs to be graduated to this standard, but at least part of it should. The width, length and thickness of the instrument is a matter of personal choice. Most engineers carry a short rule in their pocket. When a rule is used for marking out it is usually considerably longer, normally 1 ft (300 mm). However rulers of double and treble that length are available. It is never

Three pairs of calipers. On the right is a standard outside caliper with a firm joint. In the centre is a pair of sprung inside calipers—sprung calipers will give a greater degree of accuracy than firm joint ones. On the left is a pair of firm joint inside calipers which have had the jaws reversed to enable them to take outside measurements.

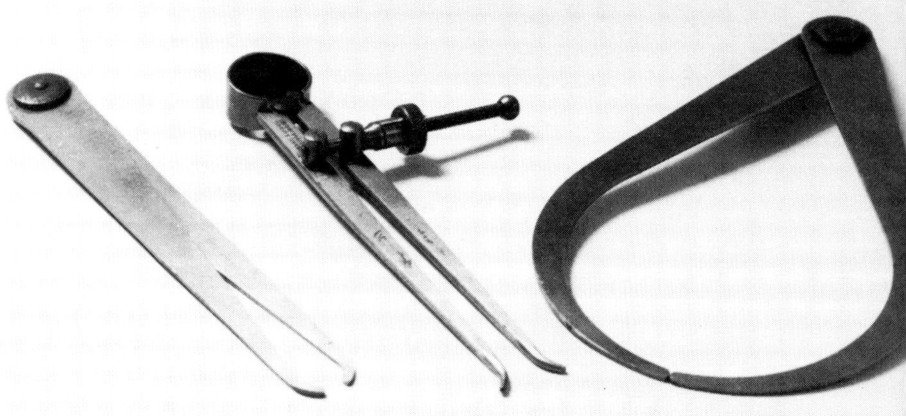

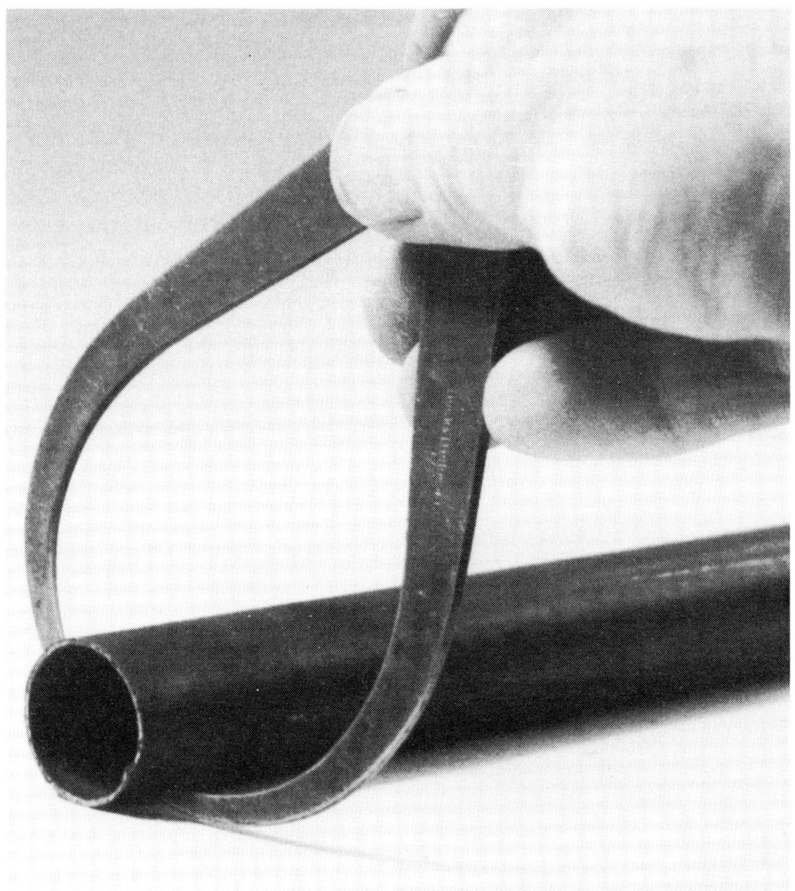

The outside caliper is used for measuring objects. The picture shows them in use on a piece of tubing. They are adjusted until they just touch the largest diameter on the tube, then taken away and the distance between the points measured. Inside calipers are used in the same way and in this case could be used to measure the inside of the tube.

worth while buying cheap metal rulers as they may not be accurate, so pay for a decent one and it will repay you many times over. They do not wear out, so it is yours for a lifetime.

CALIPERS

There are several types of calipers available, and I suppose it is fair to say that their use in marking out is somewhat limited but they do serve some purpose which is why they are mentioned here. Outside calipers are used mainly for measuring the outside diameter of round items, although there is no reason why they should not be used for flat or square items as well. They are set by allowing the ends of the arms to just touch lightly on the edge of the surface being measured. The fit should be such that both sides can be felt to be touching, whilst there is no drag when they are pulled away. The gap between arms is then measured with a ruler or a vernier gauge. Inside calipers do the same job but as the name implies, measure the insides of objects, their actual operation is the same as the outside version and again any measurement must be obtained from a ruler or other measuring instrument. These two types of calipers are not particularly useful for marking out. Odd Leg Calipers, or Jennies as they are known, have one arm similar to an outside caliper and one straight arm with a point on it which is sometimes adjustable in length. The point acts as a scriber, and they are used by drawing the caliper arm along the work edge and allowing the scriber to make a mark at the place required, the correct measurement having been set before use. They can be used equally well on round or square work. The only disadvantage is that if they are not held perfectly square the measurement changes. In spite of this I find that they are one of the most useful of all marking-out instruments.

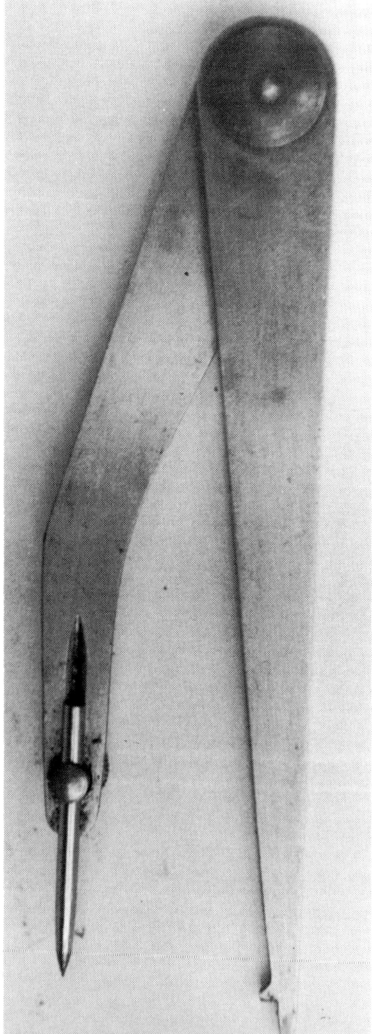

Odd leg calipers have a caliper leg and a scriber leg. They work on the principal of drawing the caliper leg along the work edge, allowing the scriber part to scribe a line parallel to the caliper leg.

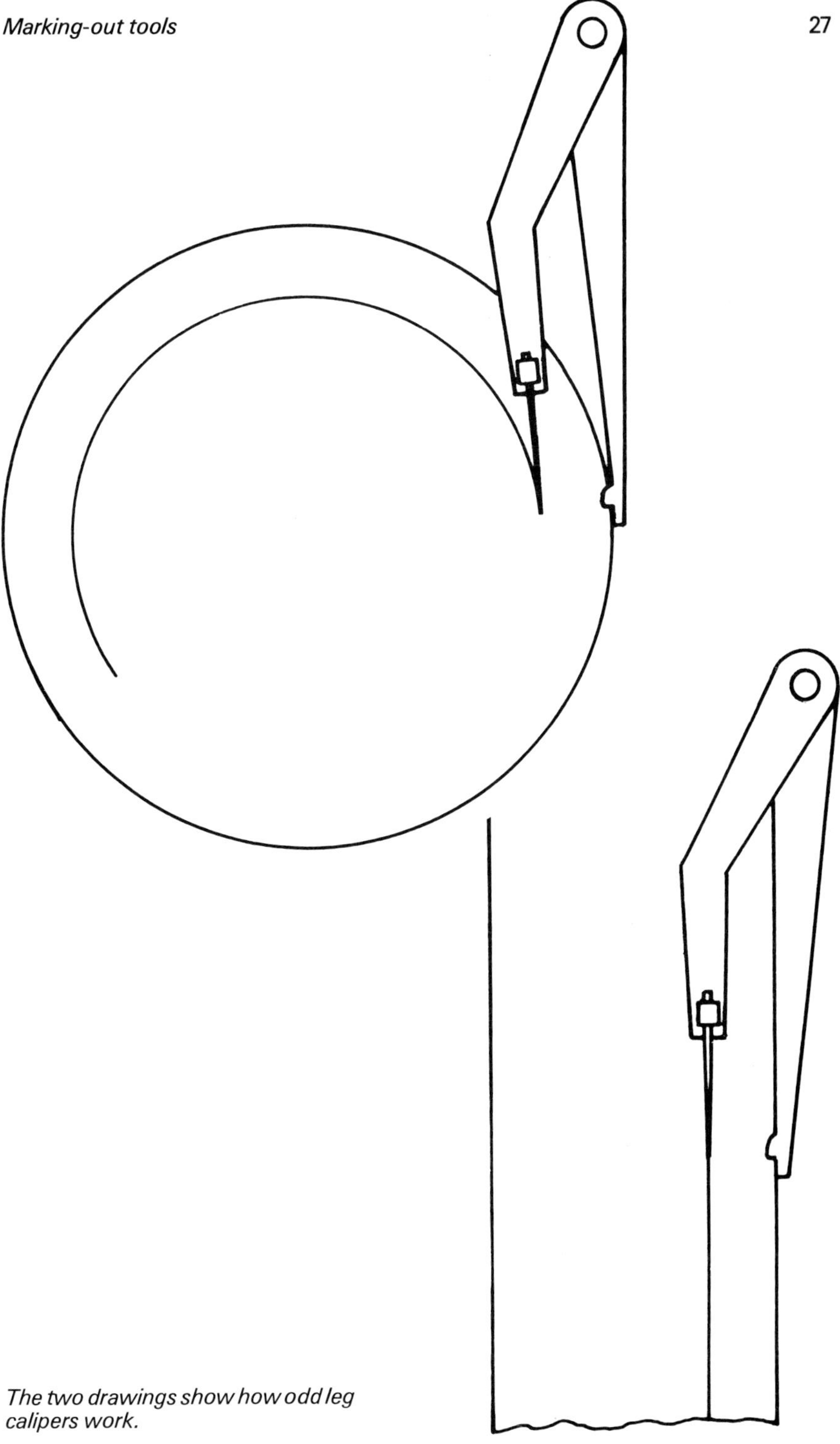

The two drawings show how odd leg calipers work.

DIVIDERS

These are rather like two scribers joined together. They are essential for marking-out purposes one point should, if possible, be put in a tiny dot scribing circular lines and for 'picking up' measurements from rulers and micrometers etc, to be transferred to the work. When using them for marking out purposes one point should, if possible, be put in a tiny dot punch mark which will prevent movement of the dividers.

Left—*A pair of dividers is like two scribers joined at the top. They are used to scribe circles and to get accurate measurements on work.*

Below—*This is a trammel. It serves the same purpose as a pair of dividers, but will open to a much larger size. The screw on the right makes accurate adjustment easy.*

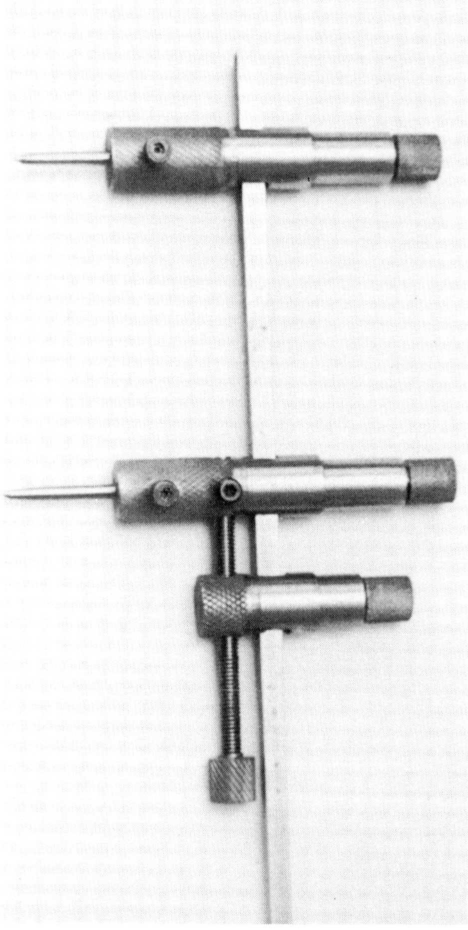

Engineer's squares.

SQUARES

Most people have some idea of what a square is like and although they tend to get called a variety of names, engineer's square, try square or set square, they basically consist of two pieces of steel set at a right angle to each other. They are set by the manufacturers and should be accurate. To check this put the square on a piece of wood or the edge of a bench with the thick part to the right and draw a line along the square at 90°. Turn the square over so that the thick part is to the left and draw a line about a millimetre away from the one you have just drawn. If they are parallel the square is right, if there is any deviation it is not accurate and will be of no use whatever. It is also possible to get what is known as combination sets. These consist of a ruler with a piece that will fit on to make a square, which is held to the ruler with a screw fitting. Other attachments include sometimes a protractor, and possibly a means of drawing angles at 45°. These are all very well but they cannot possibly be 100 per cent accurate. No matter how good the product, the screw fitting will eventually wear and cause problems. Likewise the protractor will only be as accurate as are the thickness of its lines and there is no guarantee that the angle will be right. Even so combination sets do have their uses and not to be ignored. Another tool in this range is an engineer's bevel, it is similar to a square but has an adjustable blade. By constructing an angle by geometric means, described later in the book, a bevel gauge can be set extremely accurately indeed. To sum up, a good square is a must, the others are desirable luxuries.

Below—*A combination set consists of a ruler, with attachments, one which will give a forty five degree angle, one that makes a square, and another that is an adjustable protractor.*

Above—*An engineer's bevel consists of a bar in another slightly heavier one. It can be adjusted to any angle as required. In practice the angle is worked out on the metal and the bevel adjusted to fit it. The angle can then be repeated as required.*

PRECISION MEASURING INSTRUMENTS

There are two types of precision measuring instruments, micrometers and vernier calipers. Both are available in either imperial or metric measurements and in a variety of sizes. Both are expensive but desirable items. If expense is a problem, fine models can be made without the use of either simply by being extremely careful in measuring with dividers and rulers.

Micrometers work on the principal of winding a screw up and down—by measuring how far the screw travels the instrument is able to give a reading showing the distance between the bottom of the screw and the base. Of course in reality it is a little more sophisticated than this, but that is the basic principal.

It is necessary to learn how to read a micrometer and some guidance will be found helpful. There are two sets of measurements on a micrometer, one running along the length of the barrel and the other marked around the part which turns which is called the thimble. In the case of an imperial instrument the barrel is divided into tenths of an inch on the top line of the scale, which can be further subdivided by four using the bottom line to give a measurement to the nearest fortieth of an inch, the measurement being read off from the point where the edge of the thimble rests. In order to obtain a more accurate reading the thimble is divided into 25 equal sections which allows measurements to be made in thousandths of an inch. On a metric micrometer the markings on the barrel are in millimetres on the top line and half millimetres underneath. There are 50 gradations on the thimble which allows readings to be made to the nearest hundredth of a millimetre.

A micrometer.

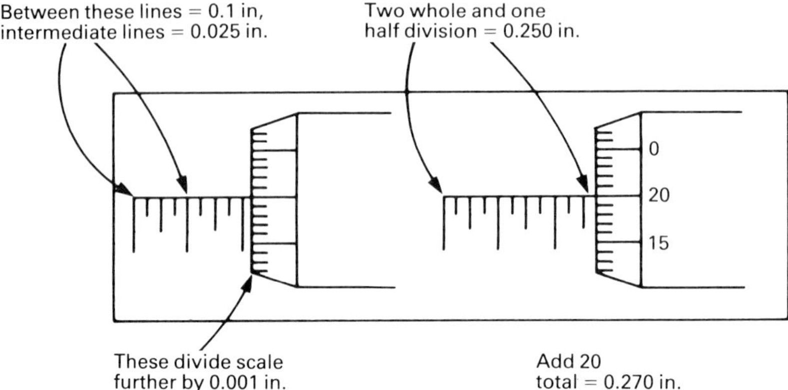
Between these lines = 0.1 in,
intermediate lines = 0.025 in.
Two whole and one
half division = 0.250 in.
0
20
15
These divide scale
further by 0.001 in.
Add 20
total = 0.270 in.

Above—To read an imperial micrometer

The scale on the sleeve or barrel is divided into tenths of an inch, and these tenths are sub-divided into four. This means that the large divisions represent 0.1 in and the four sub-divisions each measure 0.025 in. If the zero on the thimble meets at a point just after the first large division, on the first of the smaller divisions we have a reading of 0.125 in. The second small division would give us 0.150 in, the third 0.175 in, and so on. This measurement can be further sub-divided by the 25 divisions on the thimble. If then, instead of being on zero, the thimble reading was 14, we would now have readings of 0.139, 0.164 and 0.189. In other words 0.014 would be added to the original measurements.

Below—To read metric micrometer

The scale on the sleeve or barrel is divided into millimetres and half millimetres. Unlike the imperial micrometer, these two measurements usually are placed above and below a line, for convenience. Which way round they are will depend on the make of the micrometer. This makes reading somewhat easier than with the imperial micrometer as we now have a simple case of just counting the number of large divisions and adding just one small division if need be. Let us say then that the thimble reading is zero and seven whole divisions and one half are exposed, the reading therefore is 7.5 mm. The thimble is divided into 50 parts, giving us readings of one hundredth of a millimetre. If then instead of being at zero the thimble had been at 32 the reading would have been 7.82 mm. If the half millimetre division had not been uncovered then the reading would have been 7.32 mm.

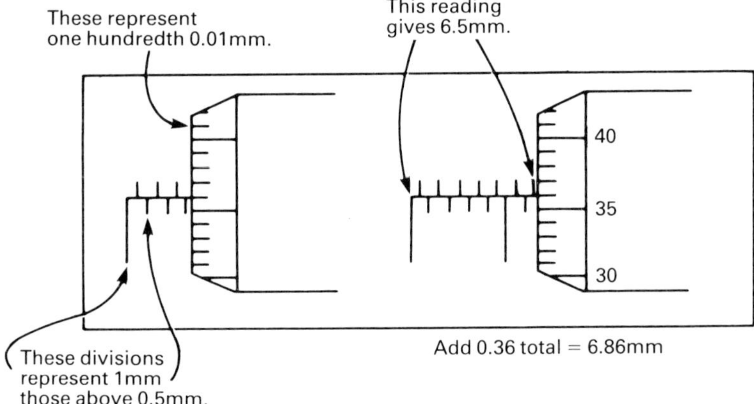
These represent
one hundredth 0.01mm.
This reading
gives 6.5mm.
40
35
30
These divisions
represent 1mm
those above 0.5mm.
Add 0.36 total = 6.86mm

Reading vernier calipers, whether metric or imperial, is rather like reading a micrometer. The difference is that the small scale graduations are spaced along the sliding centre part of the instrument. When reading a micrometer the reading is taken as being the one nearest to zero on the barrel, whilst in the case of the vernier the reading is the one that meets the nearest line. When the reading falls between two divisions it sometimes takes a good long look to decide which line is nearest, so if you need glasses for reading make sure that you wear them when using this tool. An advantage of a vernier is that it can usually measure inside diameters as well as depths of holes. It will also measure a variety of lengths whereas a micrometer will only measure a span of 1 in and a variety of sizes of tool or special adapters are required to measure items of different lengths.

Vernier gauges are available with a dial reading which takes the place of the small scale. It is easier on the eyesight and so I suppose better for accuracy. Inside and outside calipers, as well as dividers can be bought with a dial gauge incorporated. This enables one to get a more accurate measurement.

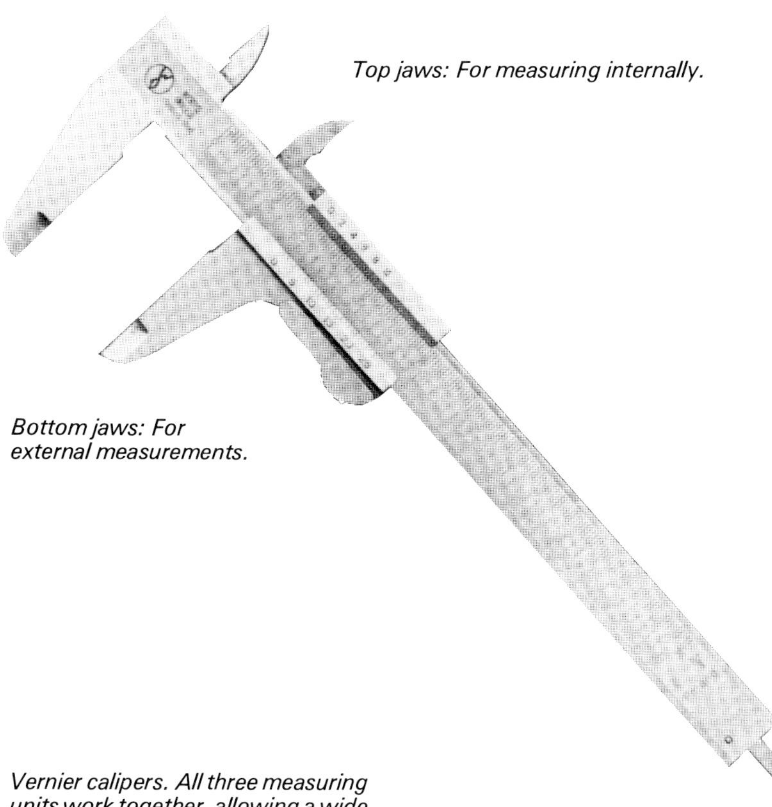

Top jaws: For measuring internally.

Bottom jaws: For external measurements.

Vernier calipers. All three measuring units work together, allowing a wide variety of measurements to be obtained.

Depth gauge.

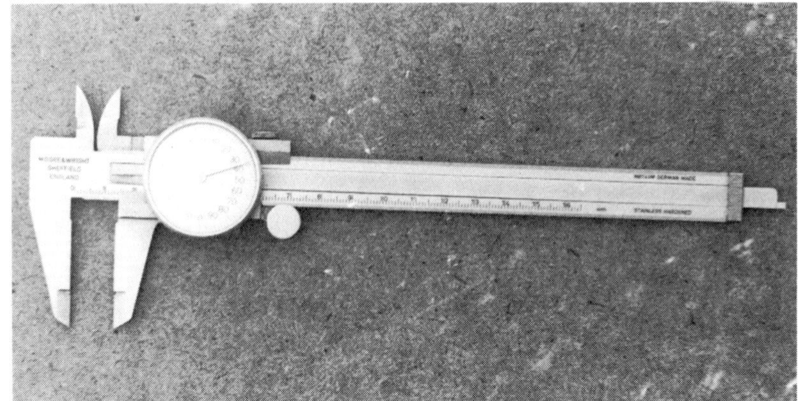

Above—*Dial vernier calipers. These work on the same principal as the ordinary vernier, except that the lowest fractional reading is taken from the dial instead of the scale.*

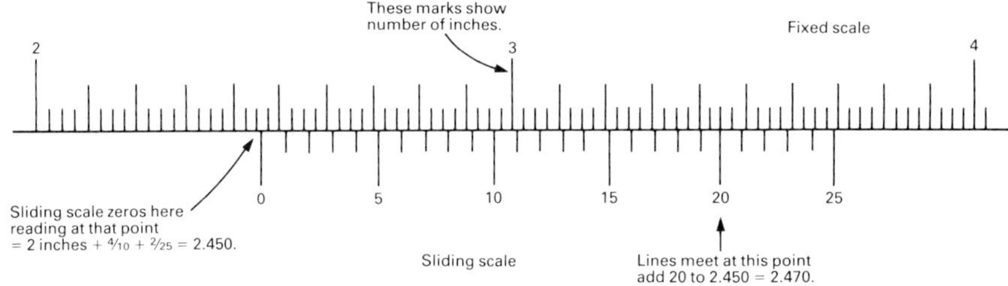
These marks show number of inches.
Fixed scale
2
3
4
0
5
10
15
20
25
Sliding scale zeros here reading at that point = 2 inches + 4/10 + 2/25 = 2.450.
Sliding scale
Lines meet at this point add 20 to 2.450 = 2.470.

Above—To read an imperial vernier scale

The fixed scale is the same as in the case of a micrometer, tenths, divided into four. The sliding scale is divided into 25 as is the thimble on the micrometer. The reading is obtained by seeing which of the divisions on the sliding scale is exactly opposite a line on the fixed scale—sometimes a magnifying glass will be needed to decide this. The one that is, is the one giving the required reading.

Below—To read a metric vernier scale

The fixed scale will contain whole and half millimetres as on the micrometer. The moving scale will have a reading of 50, again parallel to the readings on the thimble on the micrometer. The reading that is adjacent to a line on the fixed scale is the one taken.
In the illustration, for the sake of clarity, the sliding scale has divisions in 1/50 mm instead of 1/100 mm so there are only 25 divisions instead of 50.

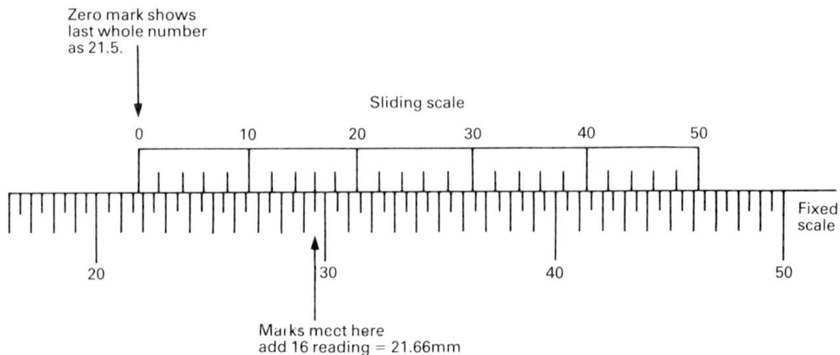
Zero mark shows last whole number as 21.5.
Sliding scale
0
10
20
30
40
50
Fixed scale
20
30
40
50
Marks meet here add 16 reading = 21.66mm

SURFACE PLATES

It is advisable to have an absolutely true surface on which to mark out. Surface plates are a large flat piece of cast iron which have been ground or scraped to a very fine accurate finish to give such a surface. They are available in three grades, each of which has a different level of accuracy. Frequently a handle is fitted to the ends to enable them to be moved. They are very expensive and very heavy and in the case of most model engineers I am quite sure that they are out of the question for these reasons. The alternative is a sizeable piece of plate glass. Plate glass is manufactured to remarkably high standards of accuracy and the flatness will be more than sufficient for this purpose. It can frequently be bought in the sort of size required, say 2 ft × 1 ft, very cheaply. It is not so heavy to shift about and will do very nicely. The best of both worlds I suppose would be a piece of plate glass for large work and a small surface plate for the smaller work. I have a little surface plate 9 in × 6 in and it is just right for the smaller marking-out jobs. For work to be marked out whilst laying flat, a piece of board with a steel strip fixed to it will work quite well, and the use of this will also be described later.

A suitable support will be needed for marking out. If a surface plate is not available, or too small for the work in hand, then a piece of flat board with a bar screwed to it will be almost as good.

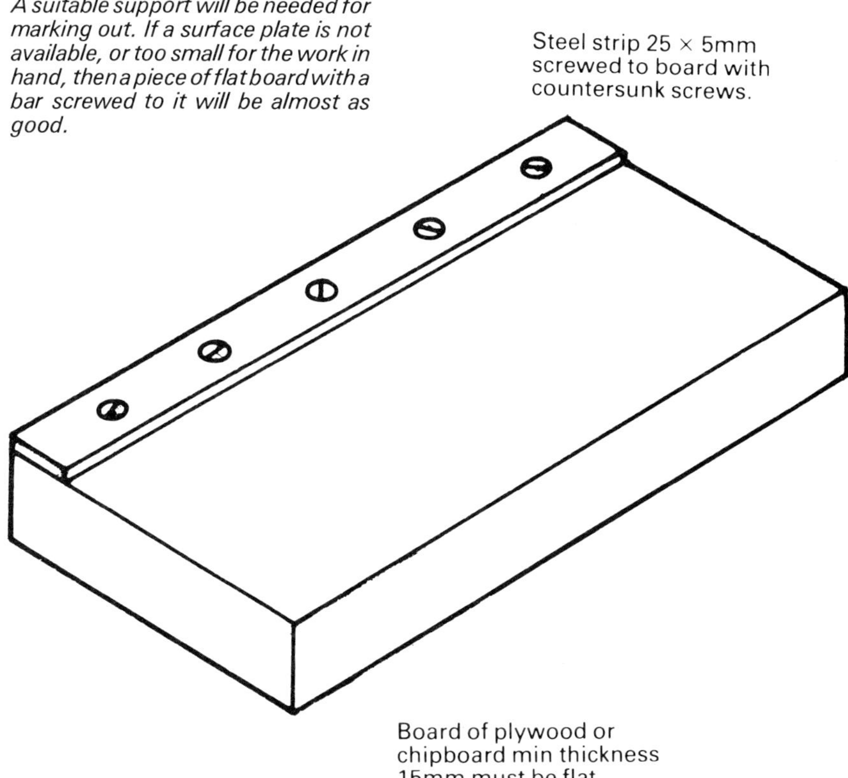
Steel strip 25 × 5mm
screwed to board with
countersunk screws.
Board of plywood or
chipboard min thickness
15mm must be flat.

OTHER ITEMS

In conjunction with the surface plate or other level surface two or three other things are available; an angle plate, which is a heavy cast iron angle that has been machined to a reasonable accuracy, and vee blocks. These as the name implies are cast iron, or steel blocks with deep vees machined in them. They are usually used for holding round work. Something else used with a surface plate is the scribing block. The chapter on workshop drawings showed roughly what one of these looked like. It is just a base with a pillar and a means of holding a scriber to that pillar and making it adjustable. A refined version of this is a vernier height gauge. This is a vernier on a base, and fitted to it is a flat scriber. It is a way of getting very accurate work. Unfortunately buying one is also a way of getting rid of a lot of hard earned money—they are very expensive and I doubt if the expense would be justified for the majority of model engineers.

Well the description of the tools for marking out has gone on a bit, and there are some that I haven't included. It all sounds very frightening doesn't it? Don't worry the next chapter will explain how you can achieve very accurate marking out with nothing more than a rule, square, scriber, dot and centre punches and a pair of dividers. I will explain the use of the other tools, but you can manage perfectly well without them.

An angle plate, seen here standing on a surface plate, is useful for holding work whilst marking out is in progress.

Above—*A vee block is used to support round work whilst marking out.*

Below left—*A simple scribing block consists of an adjustable scriber on a pillar supported in a metal base.*

Below right—*A slightly more advanced scribing block. This has a means of adjusting the scriber very accurately. There are also retractable pins in the base to allow the block to be drawn accurately along the edge of a surface plate.*

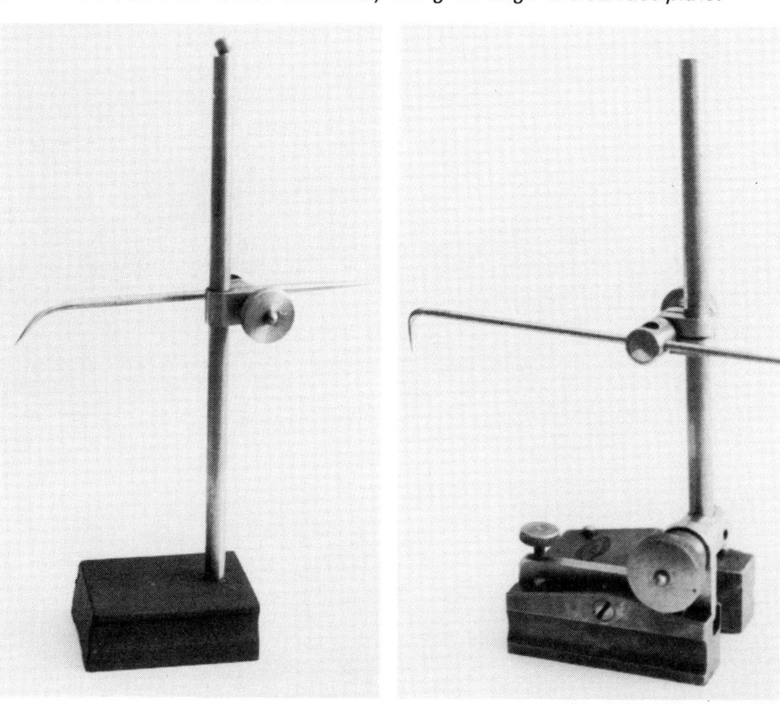

4 *MARKING OUT FLAT MATERIAL*

Having described the tools required, now let us think about how they are to be used. It would be easy enough to describe various situations and what to do when these arise, but I feel, personally, that an exercise in actually doing the job will be better. I have therefore made a drawing of what amounts to the front end of a locomotive frame. This covers most operations that will be needed where flat metal is concerned, and once the method has been grasped then it can be adapted to most other things. The frames consist of flat steel, 3 mm in thickness, 75 mm wide and they would, if shown in full length, be approximately 600 mm long. As they are steel, before starting you would need to coat them with something that will enable you to see the marking out when it is completed. If using brass or copper this probably would not be necessary, but marks on steel tend to be a bit difficult to see.

Below—*Drawing of part of the frame of a locomotive to be used as an exercise in marking out.*

Below right—*If no surface plate is available, a marking out board will have to be used. Set the metal and ruler flush against the metal strip to ensure that accuracy is obtained.*

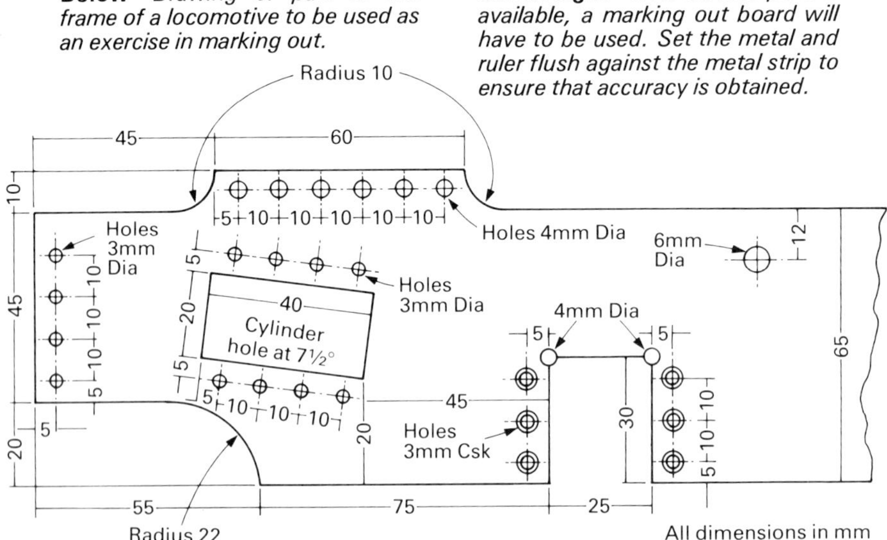

MARKING-OUT FLUIDS

Marking-out fluid can be purchased at any good tool stockist. It consists of a dye in a quick drying spirit based liquid. It comes in a variety of colours and can be bought in the form of a spray or in bottles, to be painted on with a brush. First the metal must have surplus grease and oil removed and then it is painted over and left to dry. All that sounds fine, but these marking out fluids are not as permanent as one would expect. They tend to come off with a bit of rubbing, and will certainly come off with a drop of oil. An alternative is to use felt tipped pens. These are useful for small work, not so good for the job we have in mind, and they too tend to wear off. The old fashioned method was to thoroughly clean the metal, then drop a couple of teaspoons of copper sulphate crystals in a jam jar half full of water, and stir till they dissolve. This liquid was then painted on the metal which turned it a pleasant copper colour, any scribing marks showing through as a silver colour. This coating will not wash off and old fashioned as it might be, I personally do not think it can be beaten. One other idea is to spray the steel with paint, a nice quick drying one, and this too will not rub off, and the marks will show through. The only snag is that it is not too easy to remove afterwards, but if the finished work is to be painted and a primer is used to mark on, it can be useful.

SETTING THE WORK UP

The piece of metal will obviously need to be of the right thickness, of a suitable length and a suitable width. Longer and wider than required will do no harm within reason. If we look at the drawing there are two sections cut out at the top, one on the left hand side, at the bottom, one for the hornblocks which hold the axleboxes at 130 mm from the left. There are various holes, and about as awkward as it could possibly be, a hole for the cylinders set at an angle of 7½°. There are also two different types of curves each with its own radius.

Start by checking that the metal has the left hand edge perfectly square. This is done by putting a square along the bottom edge and checking for daylight along the left hand edge. If it is right, or when it has been made right, you must decide on how the frames are to be marked. Ideally they should be put on a surface plate or piece of plate glass, against an angle plate and worked on in an upright position. Many people, and I am one of them, do not have a surface plate of the right

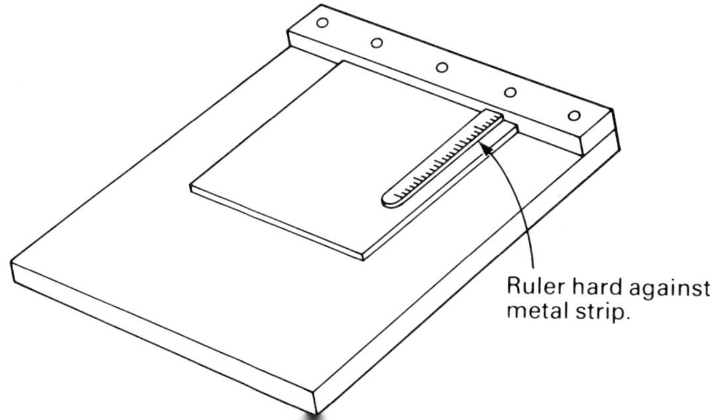
Ruler hard against
metal strip.

size. I use plate glass and the work is none the worse for it. Do not be tempted to rest the work this way up on a bench or piece of board. It just will not be possible to get it right. If a suitable surface is not available then it will have to be marked out whilst flat. This can be done on a bench, but a proper marking-out board is easy to make and the metal strip (check that it is flat) can be used to support the work.

SEQUENCE OF OPERATIONS

The horizontal lines should be marked out first. If you are marking out on a surface plate or plate glass then use a scribing block or, if you are rich enough, a vernier height gauge. Scribe the lines with one movement only. Never try to repeat a line as it will not usually work—the first mark must be the only one. If you are working flat there are two ways of doing the job. The first is to measure the metal at each end to the required point. These measurements can be gained easily enough, by placing both work and ruler against the edge on the marking-out board and just making a small mark at the required point. The two measured points are then joined with a scribed line, using a straight edge and scriber. It will probably be fairly obvious that this is a somewhat hit and miss way, as it relies entirely on the eye. A small gauge can easily be made that will ensure a straight, accurate line and it will repay its worth over and over again. Another way is to put the flat edge of a pair of odd leg calipers against the edge of the metal after they have been set to the required dimension. Using the scriber draw them along the metal. Once again accuracy will depend on how steady you are, and if the 'Jennies' twist the dimension will alter. So really the gauge is best, as the flat edge gives

Above—*When the height of the scriber has been set the metal is clamped to an angle plate and the scribing block slid along the surface plate, with the scriber in contact with the metal.*

Below left—*To set the height of the scriber in the scribing block a ruler is stood on end when using a surface plate, and the scriber matched with that. The ruler can either be clamped to an angle plate or, as here, a special block can be made to hold it.*

Below—*For drawing lines parallel to an edge a simple gauge like this can be made up. It can also be fitted with a blade and used for cutting thin sheet metal.*

Bottom—*To set odd leg calipers, put the caliper edge on the end of the rule and set point to mark required. Note the use of an engineer's rule which is accurate at 20°C, at any other temperature there could be some very slight error due to expansion.*

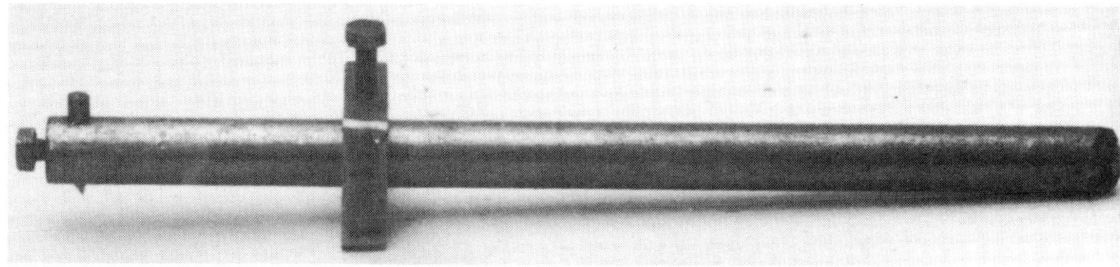

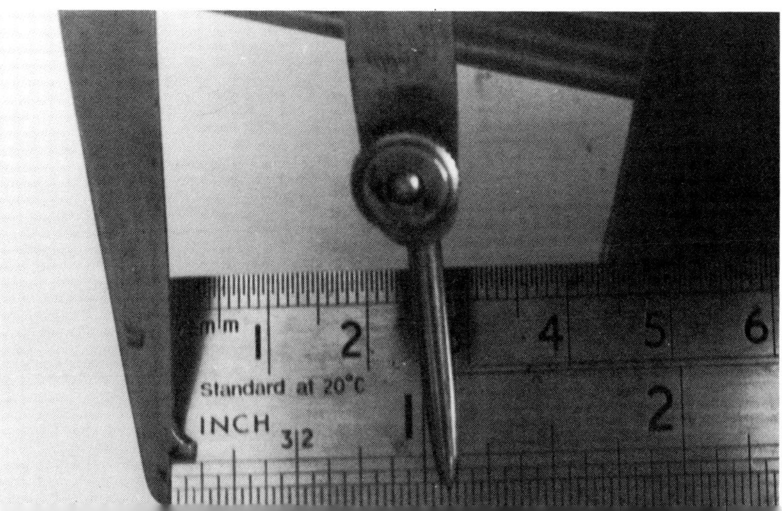

greater purchase on the metal. Include in the horizontal lines, all centre lines for rows of holes on a horizontal plane, but of course your line will be a plain scribed one rather than a dot and a dash as shown on the drawing.

DATUM POINTS

Next you will need to put in the vertical lines. Again these include any for rows of holes, and in the case of the frames we will need also to put in small vertical lines about 6 mm to 7 mm long where the ends of the radiuses will come. It will probably be as well to make the vertical lines by using a square and a scriber, although there is no reason why the frames should not be turned on their end on the surface plate and the scribing block used.

At this stage a couple of words about the measurements would not go amiss. All measurements should be taken from the base or left hand datums. If we take the base, working from the left we see a measurement of 55 and next to it one of 75, followed by 25. These should not be measured in this way. It should be 55—130—155 and so on. If you use the measurements as they are shown then if each one is accidently measured ½ mm over length, and there are twenty measurements, there will be an error of 10 mm at the end and the whole thing will be useless. The method I suggest leaves far less opportunity for errors. Measurements should be 'picked up' from the ruler or vernier gauge by means of dividers or odd legs. Do not rely on putting the scriber adjacent to the required mark on a ruler as this again can lead to considerable error. If the measurements are too long for dividers then the correct tool to use is a trammel, however if you do not have one then you will have no choice but to use the rule and scriber method. If this is the case, check the measurements and marks several times before making the next one.

Setting dividers to a rule measurement. Place one point at the mark required, plus 10 mm or 1 in. Then adjust until other point settles into the 10 mm or 1 in mark. Do not attempt to use ruler end with dividers.

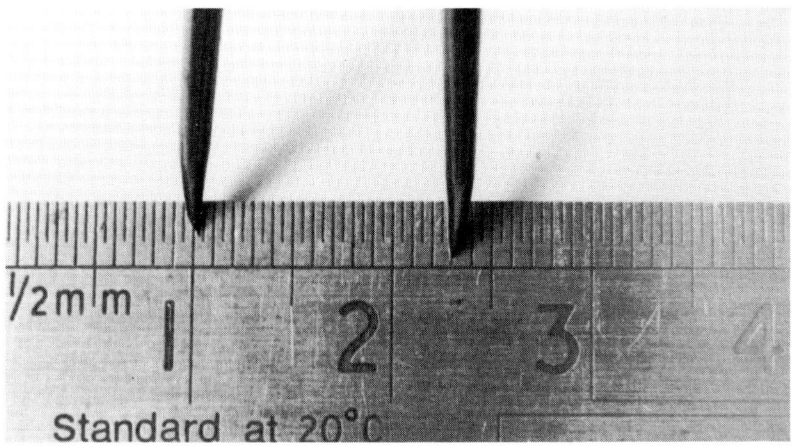

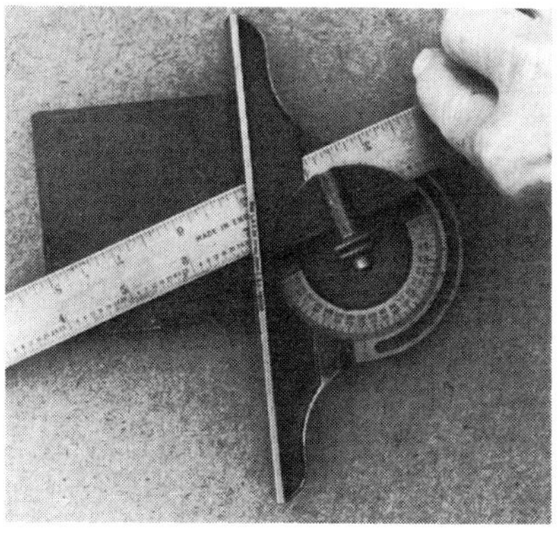

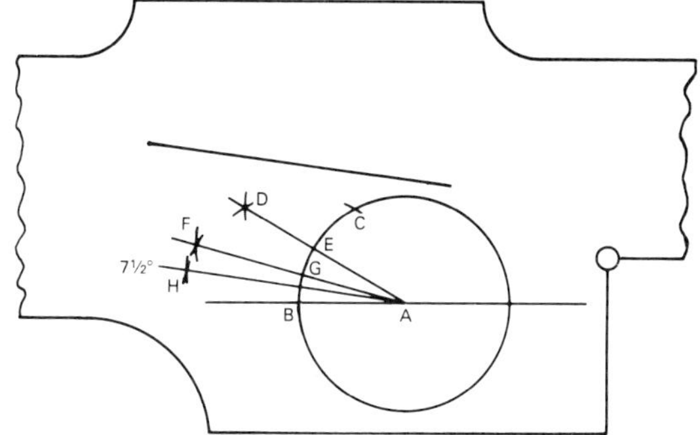

Top—*To draw parallel lines using dividers put point of divider at A and make arc above. Repeat at B. Connect tops of arcs with line.*

Above left—*A combination set used to obtain an angle.*

Above—*How to obtain 7½° by simple geometry.*

ANGLES

The next task is to mark out the cylinder hole which lies at an angle of 7½°. This is somewhat awkward and whichever method you use the point marked A is going to be of great importance. If you have an engineer's protractor, then the angle can easily be found. Such equipment is expensive and not entirely a necessity, the same result can be obtained from the combination set, except that it is more difficult to work to half a degree. It is quite possible to find this angle and several others by means of simple geometry. If you take a pair of dividers, scribe a circle and then put the point of the dividers on to the circumference of the circle, without changing the setting, make a mark, and then from that mark make another and so on, it is possible to divide the circumference up into six equal parts. If a line is then scribed from each of these marks to the centre you have six angles of 60°. By further division of those angles, using the method below, you can obtain an angle of 7½°.

Start by scribing a line horizontally to cross the point A. Make a tiny pop mark where it intersects. Set the dividers to a diameter of about 30 mm, and using the pop mark scribe half a circle at the top of the line. Without changing the setting, make an arc on that circle from where the circle meets the horizontal line. It is marked point B. Call the mark you have just made C. Centre pop lightly B and C and from each one make arcs at D. Centre pop D and draw a line through to A which will intersect the circle at E. Using B and E make two further arcs at F and scribe a line to A from there, calling where it meets the circle G. Using G and B repeat the operation and from where the arc meets, point H, scribe a line to A. This will be at 7½°. The angles so constructed being 60°, 30°, 15° and 7½°, you now have the bottom line of the cylinder hole marked.

To get the two rows of holes and the other parallel line for the cylinders set the dividers to the required measurement, make two arcs, one from each end of the line you have constructed, connect them with another line and you have the line running parallel. If an engineer's bevel is available it could be set to the original constructed line and used in that way. The end lines will have to be at right angles to the top and bottom ones. In order to do this you will need to mark two points along the original line at equal distances from A. Let us call these I and J. Set the dividers to the distance between points I and J and scribe an arc both above and below the line from each of these points. Scribe a line

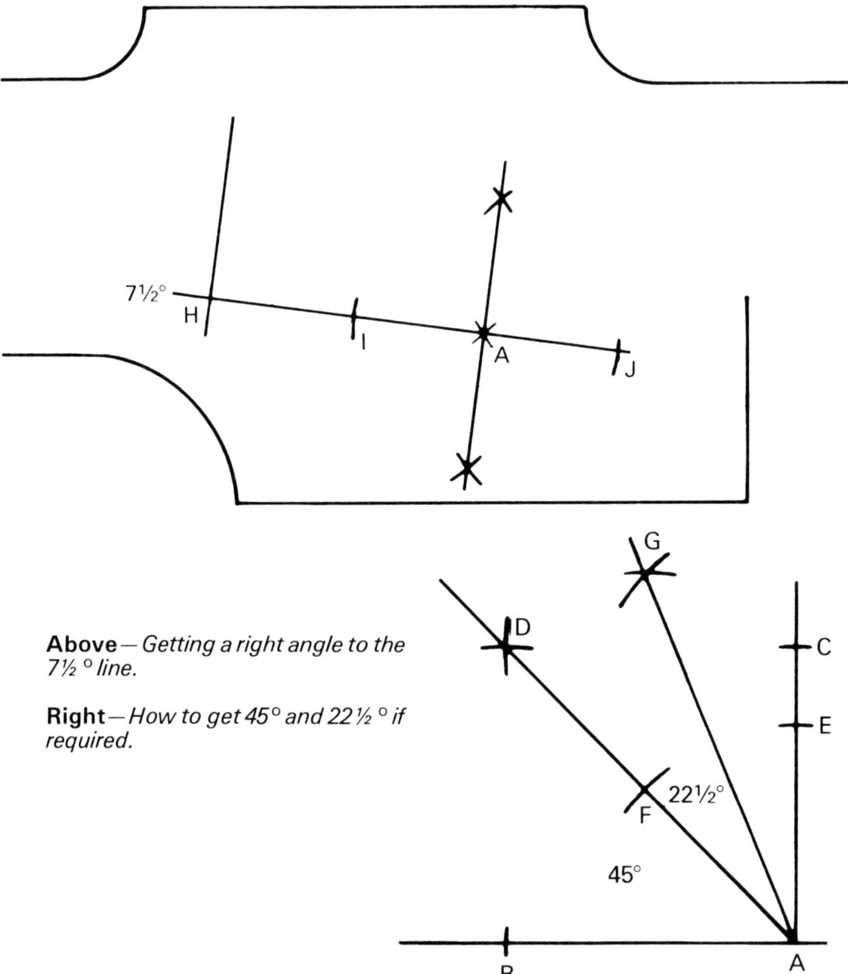

Above—*Getting a right angle to the 7½° line.*

Right—*How to get 45° and 22½° if required.*

between the two points where the arcs cross. This will pass through A at right angles to the original line. To mark out the left hand end of the cylinder hole use the same technique at the other end of the line.

You now have reached the point where all the lines have been scribed. The next thing is to mark out the holes. This is done by scribing little marks across those that are there already. They are then marked with a centre punch, or perhaps I should say a dot punch, the centre punch coming later. It is possible to feel with the punch where the lines intersect and it should be tapped very lightly with a hammer first of all. If for any reason it is not quite right, lay the punch at an angle. Tap lightly and correct the error. When it is absolutely right use the centre punch to make a heavier mark. More detailed instructions will be given on marking out for drilling holes in order to get accuracy in Chapter 9.

When the holes are all marked out, mark lightly along the lines to be cut with a dot punch. This makes them show up better when cutting and filing the work. The cylinder hole now needs lines drawn inside it as shown on the drawing. They are spaced 2 mm from the cutting lines. At 3 mm intervals dot punch right round all four lines. Increase the marks with a centre punch. These can then be drilled with a 3 mm drill. The holes will be just short of the cutting line, which will allow for filing up, and they will just meet each other sideways. When you have finished the drilling it will be possible to break the piece of metal out, leaving the hole to be cleaned up with a file.

The only part not covered so far is the cut out for the axleboxes, or hornblocks. The vertical lines are straightforward enough, but the horizontal one might need to be dealt with in the same way as the cylinder block hole. The two holes at the corners are there to assist cutting out, and it will be possible to cut lengthways from these, with care, using an abrafile or coping saw.

The frame in the drawing at the beginning of this chapter marked out.

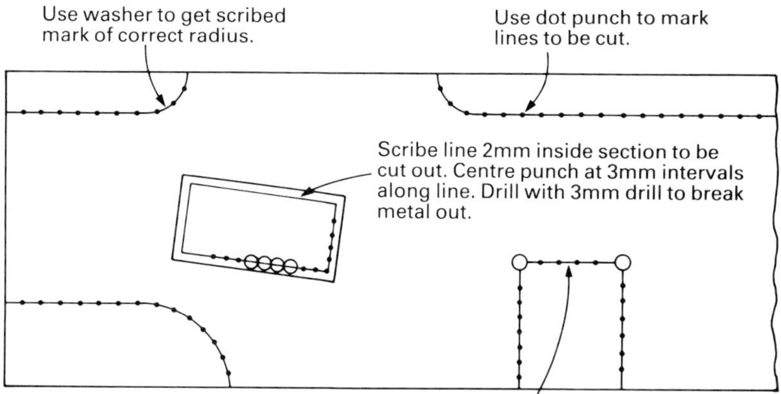

Use washer to get scribed
mark of correct radius.
Use dot punch to mark
lines to be cut.
Scribe line 2mm inside section to be
cut out. Centre punch at 3mm intervals
along line. Drill with 3mm drill to break
metal out.
The top can either be cut as per the cylinder hole
or by using holes in corners cut along the line with
abrafile or coping saw.

5 *MARKING OUT ROUND AND IRREGULAR WORK*

The principal of marking out work that is of round or irregular shapes is much the same as for marking out flat sheet. The only problems are the fact that it is often not possible to hold the marking out tools firmly against the work. A few tips on how this can be overcome will be given, but I am afraid that when it comes to things like castings it is a case of using your initiative. The principals explained will without doubt help somewhat, and they must be adapted to whatever the shape is that is being worked.

ROUND WORK

Round work is not too difficult. It should always be stood in a vee block

for support, and if one with a clamp is available so much the better. The vee block should be stood on the surface plate or plate glass. Marking the ends of the material can be done with a scribing block, or with odd leg calipers. Once the first mark has been made put in a small dot mark and mark off from that using a pair of dividers. The scribing of horizontal marks is easy, but marks at right angles to those are much more difficult, and the only way if you have limited equipment is to turn the work and test it with a square that rests on the surface plate. When the arm of the square is exactly touching the mark right the way along its length, clamp the work tightly and then use the scribing block again. The operation calls for considerable care if it is to be accurate.

Marking the side of a round bar lengthways is just a case of scribing it by running the scribing block along the surface plate or glass. Marking round the circumference is quite difficult. Ideally it should be done with the work in a lathe. If this is not possible you will have to resort to a steady hand and sharp scriber. If the end of the bar is true and the mark required is not too far from that end, a pair of odd leg calipers can be

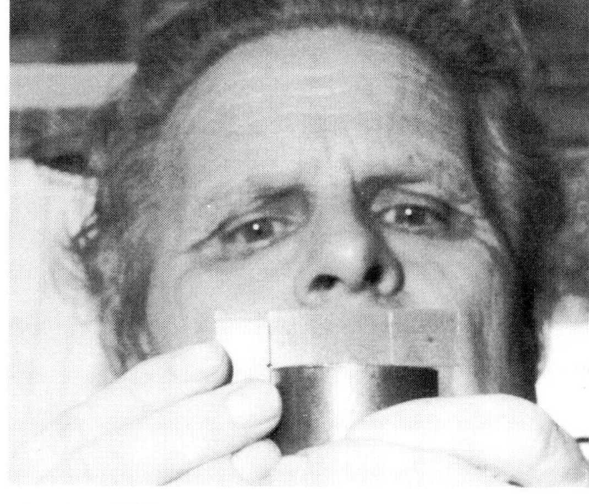

Right—*The end of a round bar can be checked with a square, by holding it against the light and seeing if any light shows through. The bar end should be even all the way along.*

Below left—*A bar of round material should be supported in vee blocks for marking out. The picture shows one so held whilst a scribing block is used to mark the end.*

Below—*A scribing block can be used to mark the side of a bar, which is held in vee blocks.*

Above—*A scribing block should not be used to scribe along the top of a bar, but can be used for getting measurements at the top. The problem with trying to use it along the length is the tendancy for the scriber to wander sideways.*

Below—*Angle iron can be used for marking along the top of a bar.*

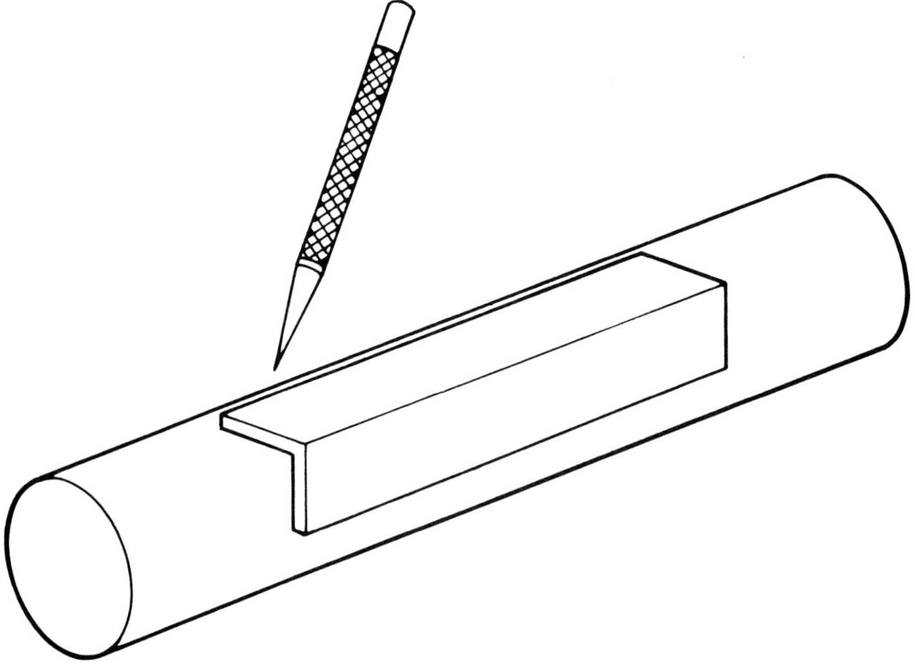

used. The caliper leg is placed on the bar end and drawn round the edge, whilst the scriber marks it as required. A little dodge that can be used is to wrap masking tape round the bar at the required place. Check every few millimetres to see that the positioning is right. When you are absolutely sure that the positioning is correct, then using a short ruler as a support, scribe the line by slowly rotating the ruler round the bar using the masking tape as a guide. The scribing block is not very successful on marks that are near the top of the bar and the best method in this case is to use a piece of angle iron to run the scriber along. The angle iron will fit firmly on the bar as it is supported on two edges.

A photograph of angle iron being used for marking purposes.

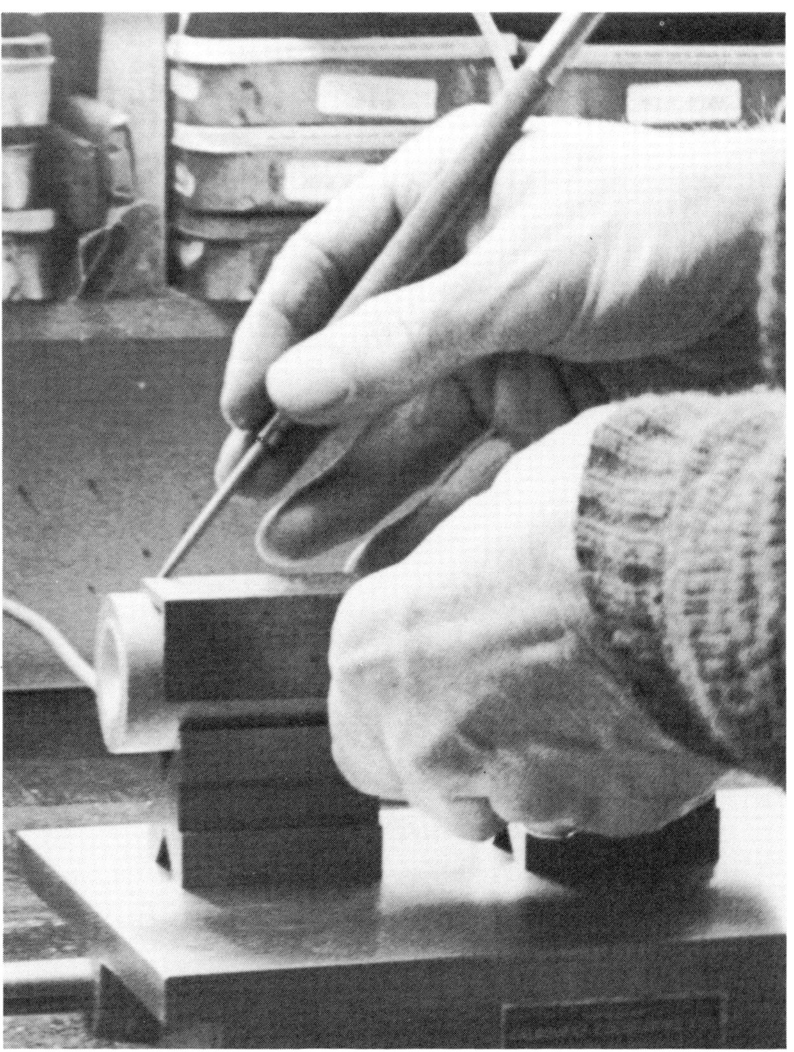

CASTINGS

The problem with marking out castings is twofold. Firstly they are all too often of irregular shapes, and secondly the castings themselves are never true. We must therefore firstly create a true surface and mark off from that. To get this approximate positioning of where that surface should be, it is marked as well as it can be and then it is machined. It must, if possible, be a flat surface. Sometimes it is as well to just mark off the next surface for machining and complete work on that before proceeding with further marking out. All main surfaces should be machined if possible before holes etc are marked off. Sometimes you will find a casting with a large hole through it, with points that must be marked at certain distances from that hole. Marking them out accurately from the edge of the hole is not possible, the hole itself being both undersize and out of true. The best way to deal with this situation is to make up a hardwood bung that will fit into the hole. Tap it in lightly with a hammer, and then file and sandpaper it flush to the castings. Any gaps that are left can be filled with plaster, and when dry rubbed smooth, so that we are now left with a complete surface from which we can mark off as we like. This idea can be adapted for use under other circumstances where there is an odd shape that it is not easy to mark from. Another idea is to clamp a steel bar across undulating surfaces and use that as a datum. Do not forget though that the thickness of the steel bar must be added to the figure that is to be used.
almost certainly means a ruined piece of work, so make sure that mistakes are not made.

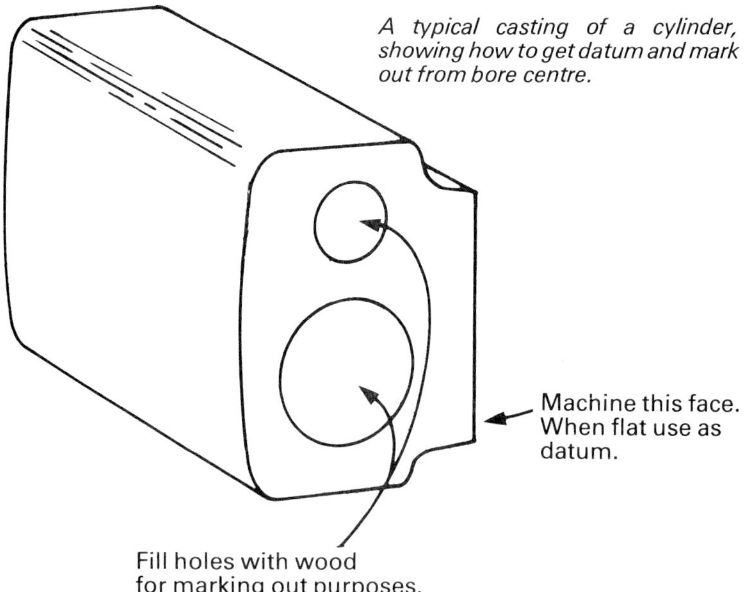
Machine this face.
When flat use as
datum.
Fill holes with wood
for marking out purposes.

A typical casting of a cylinder, showing how to get datum and mark out from bore centre.

As far as is possible within the limited space that is available I have covered the marking out of most items. If any awkward problem arises then I find that if I sit down and ponder for a while I will find a solution. All that has been written has related to marking out at the bench, as indeed it should in a book on benchwork. I would be wrong to leave it at that, and not point out that machines can provide the most reliable means of marking out that there is. Work can be mounted on the cross slide of a lathe and a scriber put in the chuck. Perfectly accurate measurements can be obtained by using the graduations on the slides, far more accurate than can be obtained using a ruler. A vertical slide gives even more versatility to the process. A milling machine too provides a good surface to which work can be clamped to be marked out, by means of a scriber set in the head. Again the slide graduations provide a means of measurement that is 100 per cent accurate. Some toolmakers, (in fact it is probably fair to say most,) rarely mark work out at all. It is set on the milling table, or perhaps on the table of a jig borer and each operation carried out according to the measurements incorporated in the machine. Such facilities are not often available to the home worker, but if possible use a mechanical means for at least some of the marking out, in order to achieve the accuracy you require. With any benchwork the watchword has got to be check and check again before making any marks on the metal at all. Only when you are absolutely certain it is right should you go ahead. Remember, a mistake at the marking out stage

6 *MATERIALS*

It is important to know something of the materials which are used in model engineering. These will be mainly metal and so that is what the chapter will concentrate on.

COPPER

Copper is a non-ferrous metal (which means it will not rust). It is obtained from an ore called pyrites. The ore is crushed and seived to exclude any useless matter. It is calcined to remove any sulphur or arsenical impurities, and then heated to remove the iron oxide. It is then smelted. Copper is a very good conductor of both heat and electricity, it is easily worked and can be rolled into thin sheets and drawn into fine wire. In the finished form it can be purchased in sheet, round, square, rectangular and hexagonal rods. It can also be purchased as tube and wire. It may be further graded into soft and hard, when purchased. The valuable properties as far as modelling is concerned is its ability to accept solders easily and to conduct heat. It is ideal for beaten metal work, for making boilers and for soldering irons. When in use it will harden with work, but can be softened by heating and then quenching in water, or by just leaving to cool. Frequently whilst working on it, it will harden and will need to be annealed (softened) several times during operations. If a sheet of copper is too soft, then run a wooden roller up and down it a few times and it will harden. Overheating can cause structural changes within the metal, mostly these appear as a crystalined form, and should therefore be avoided.

Copper is a difficult material to drill, tap and turn. Tools tend to bind, and files and drills will clog up. Work on copper therefore should always be done in easy stages. Taps, drills and dies should be frequently withdrawn from the work. Work can be assisted by using either white spirit, turpentine, paraffin, or liquid soap as a cutting agent. If the material is to be soldered or brazed after machining and the solder needs to penetrate the hole or thread, then a weak solution of the required flux should be used as a lubricant. It is difficult to clean off some cutting agents enough for soldering to be carried out, particularly if they are oil based.

BRASS
Brass is a yellow coloured metal, an alloy of copper and zinc, the quantity of each deciding the structure of the brass. Sometimes a small proportion of another ore is added to make the brass suitable for particular purposes. It is obtainable in the same forms as copper, and also in half round sections. It can be bought as hard or soft, a very hard form being known as ships, or naval brass, which is a very useful material. If the alloy contains over 65 per cent of copper it will have the unusual property of being able to bend easily when cold, but liable to fracture if bent when hot. It can be softened by heating and allowing to cool slowly. It will work harden in the same way as copper. All brasses will drill and tap easily, but if the drill is too sharp will tend to snatch as the drill breaks through. In spite of that it needs to be worked with tools with a keen edge and will not cut with hacksaw blades or files that have previously been used on steel. It solders easily and is useful for small castings. The brass usually found on screws and household fittings is more often than not of very poor quality and it is doubtful if such material should be used where strength is required.

GILDING METAL
Gilding metal is frequently used in schools, but strangely enough model engineer suppliers rarely seem to stock it. It is an alloy of 90 per cent copper and ten per cent zinc. It is particularly good for beaten metal work and solders well. It can be softened like copper and will work harden. If a workpiece has to be formed by hammering, then gilding metal is far superior to brass for this purpose. As far as I know it is only sold in sheet form.

BRONZE AND GUNMETAL
All bronzes and gunmetals are alloys of copper. Gunmetal is invariably obtained in cast form, either as castings or in cast sticks. It is particularly suitable for bearings. Phospher bronze is the most commonly used of the bronzes and is usually bought as rods and sheet. It is somewhat redder in colour than is gunmetal. Both are fairly easy to machine, but phospher bronze has a tendency towards snatching. Both are very hard wearing. Bronze, similar to phospher bronze, is available in a specially prepared form as bearings. These are impregnated with oil and need no lubrication. They can be purchased in different diameters and length. Usually they are sold as oilite bearings. Other bronzes available are manganese, which is extremely hard, and aluminium bronze. The latter is easy to machine but difficult to solder or braze, whereas all other bronzes and gunmetal will solder and braze easily.

ALUMINIUM
Aluminium is a light metal of a grey colour, which will polish to a silver colour quite easily. It is sold as rod, angle, tube or sheet, and is useful for sheet metal fabrication. It can easily be beaten into shape but is very difficult to solder, with special flux and solder being required. It is easy to machine, and will melt at quite low temperatures making it useful for small castings.

Duralamin is a name now used to cover any of the harder types of aluminium although in fact there are several other types of hard aluminium available under different names. Dural, as it is known, consists of 4.5 per cent copper, 0.5 per cent manganese and the rest aluminium. It does not solder easily, but machines well. It can be softened by heating but will reharden when left. After a day or too it will become very hard indeed. It can be cast, having a fairly low melting point, but loses some of its hardness when so dealt with.

ZINC

Although not commonly used by modellers, zinc can be a useful material. It is available in sheet form, and as such is quite good for platework. It also etches easily, making it useful for fine detail overlays. It has a low melting point but will soft solder quite easily. It is readily available at scrap merchants and can prove both cheap and useful. It is very good for making soft jaws for vices.

LEAD

Unless mixed with other metals to form an alloy, lead is not often used in model engineering. It does have its uses however. In sheet form it will make excellent soft jaws for a vice. It also has a very low melting point and can be used to make castings by simply melting in a ladle and pouring into a plaster mould. It is useful for adding weight to a model and once again the low melting point means that weights can be cast to the exact size required.

MONEL METAL

Monel metal is an alloy of copper and nickel. It is obtainable in sheet or rod form and is very hard. It is easily soldered, and as a result of the considerable strength of the material is frequently used in boiler making. It is possible to obtain pop rivets made of monel metal, and these are very useful where the use of steel ones would lead to problems of corrosion. It is not an easy metal to get, but nowadays most model engineer suppliers stock a limited amount.

NICKEL SILVER

Sometimes known as German silver it is quite hard and takes solder easily. It is obtainable in sheet or rod form, and is a very useful but somewhat expensive metal. It has similar properties to brass, but is somewhat harder and will take paint easily, whereas brass will not. It is a very nice material for making platework on models.

TIN PLATE

The name is given to steel sheet which has been tinned, this means coated with a form of solder to protect it from rust. It is fairly cheap, solders well and will bend neatly. As the name implies it comes in sheet or plate form. It will rust, as corrosion will set in where the metal is cut and the tin is not available to protect it. Food tins, petrol cans and similar objects are made of forms of tinplate and can provide a useful means of obtaining a supply of the material.

IRON

Iron is rarely seen these days in the home workshop except in the form of castings. It is excellent for this purpose and a good finish can be put on it with files etc. It is very brittle and for this reason castings are usually of thicker section than would be the case if the item was fabricated from steel. It can be purchased in cast sticks, which are very useful for bearings.

STEELS

The very word steel covers a whole world of materials. It is an alloy of iron and carbon and how hard the steel is will depend on how much carbon is in it. It can be purchased in all forms, and can be either bright or black. Black steel does not rust as easily as bright but it does not finish as well and the sections in which it is purchased are not as accurate in their measurement as are the bright ones. Black steel is fairly soft and is quite ductile. It frequently has a scale on it when purchased and this needs removing before use. Black steel angle has quite a few uses in fabrications.

Mild steel, which is what the bright mild is usually called, is available in all forms. It comes in a variety of qualities. It contains about 0.15 per cent carbon, but this may in some circumstances be as high as 0.3 per cent. From there up to 0.8 per cent carbon is classed as a medium carbon steel, but will still come within the mild steel range. If the percentage of carbon is around 0.15 per cent then the steel is usually known as free cutting and classified as EN1A. In this case, it will have mixed with it other metals such as silicon and manganese. It will also contain limited quantities of sulphur and phosphorus. Such steel is sold in bar form in a variety of shapes. It may also be precision ground. Some qualities also contain a limited amount of lead to aid machining.

Medium carbon steels are classified EN3B to EN8 and are somewhat harder than the free cutting varieties. This gives them extra strength for such things as shafts and axles, and they are particularly fine steels when purchased in the ground form. For case hardening they are far more suitable than are the free cutting ones, which because of their content do not always harden successfully.

High carbon steel is usually purchased in the form of gauge plate or silver steel. Because of the higher carbon content it is not only harder in the first place but will also harden and temper when making tools. Some steels come in the hardened form and need to be softened before use. Silver steel is bought as either round or square rod, and frequently the ends are quite hard where the steel has been cut during manufacture. The rest of the rod being soft. A recent addition to the range is free cutting silver steel that combines the best of both worlds, being easy to work whilst at the same time having the ability to harden and temper in the normal way.

STAINLESS STEEL

Stainless steel is a name that covers a wide range of steels, many of which are not really stainless at all. (This applies probably less to the purchase of metal than it does to tools that are described as stainless.) The steel is made into the stainless form by the addition of a small quantity of nickel and a small quantity of chromium. Of necessity this alters the percentage of other elements. Usually stainless steel is very hard, which is why it is commonly used for knife blades. It is now possible to purchase steels in the range that are hard but are free machining, and these are of the greatest value to the model engineer. Stainless is sold in all the usual forms of sheet and rods. Probably the hardest of the steels we will find is classed as EN58B. This has had a small quantity of titanium added, and is used where a lot of wear is likely. Of the free machining varieties EN56AM and EN58AM are probably the most popular. The most corrosion resistant of the softer varieties being EN57. Stainless steel will solder and braze quite well but special fluxes are required for the best results.

SHAPES AND SIZES

The next question must be where to purchase metal from. Model engineer's suppliers stock a wide range suitable for such needs. It is usually sold in short lengths designed to suit the model engineer. They are somewhat expensive when compared with other sources, but if one takes into account the fact that the supplier must purchase a considerable quantity, cut it to size and possibly hold it in stock for a long time, thus tying up his capital, then most of them are very fair. If larger quantities are required then go to a stockist of metal. You will have to purchase usually a stock length which may be 12 ft to 14 ft, but as a rule it will be cheaper. I say as a rule as this is not always so. The dealer I sometimes buy non ferrous metal from charges quite high rates up to a certain quantity but if I purchase over that figure then the cost comes down by over 50 per cent. It can be a good idea to go to a stockist with a friend and to divide the quantity between you. In this way you will be able to take advantage of the cheaper prices without having to buy so much. A disadvantage of going to a stockist may well be that you just go to a counter and order what you want from a salesperson who may have little knowledge of the uses of the metal sold, in which case you must be certain of what you need before you go along. A model engineer supplier will stock the material that is most suitable and so, in the long run, it might be better for you to buy there. Scrap yards can be a good source of supply for short lengths of metal, but you need to know what you want. The scrap merchant deals in ferrous and non-ferrous, and beyond that he is not usually interested, so what you get will be your selection. Factories will often sell scrap to you at remarkably cheap prices. It may be a case of buying a sackful at a time. I used to buy metal in this way, but I found myself with a large quantity of metal that I could not use. One kind factory owner used to give me a sackful whenever he thought of it. That was great until he started to use a steel that was so hard that I could not possibly use it and I had to get rid of it. Nevertheless, these people were both kind and interested and although I had some problems, I also acquired a stock of metal that I could not otherwise have had.

Metal is sold usually by weight, but the size is quoted in ordering. Round rod is quoted as diameter. Rectangular as the measurements of two sides. Tube by outside diameter and thickness, or inside diameter and thickness. Make sure when purchasing that you get the right size and that you are not quoting outside diameter whilst the seller is selling

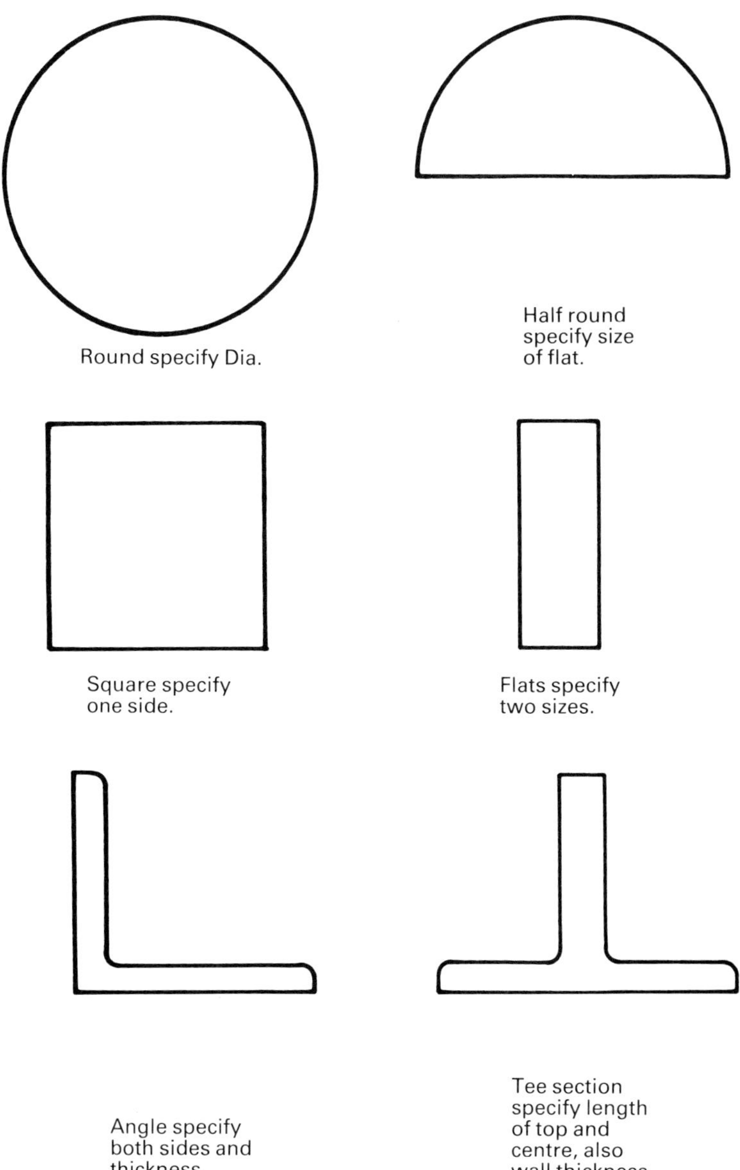
Round specify Dia.
Half round specify size of flat.
Square specify one side.
Flats specify two sizes.
Angle specify both sides and thickness.
Tee section specify length of top and centre, also wall thickness.

by inside diameter! Thickness is usually quoted in Standard Wire Gauge (SWG). A conversion table will be found for this at the end of this chapter. Sheet metal is either quoted as SWG or in thickness of the sheet in imperial or metric measurement. Angle is quoted by the size of the flats and the thickness. Tee section and girder section the same. Most materials can be purchased in imperial or metric sizes.

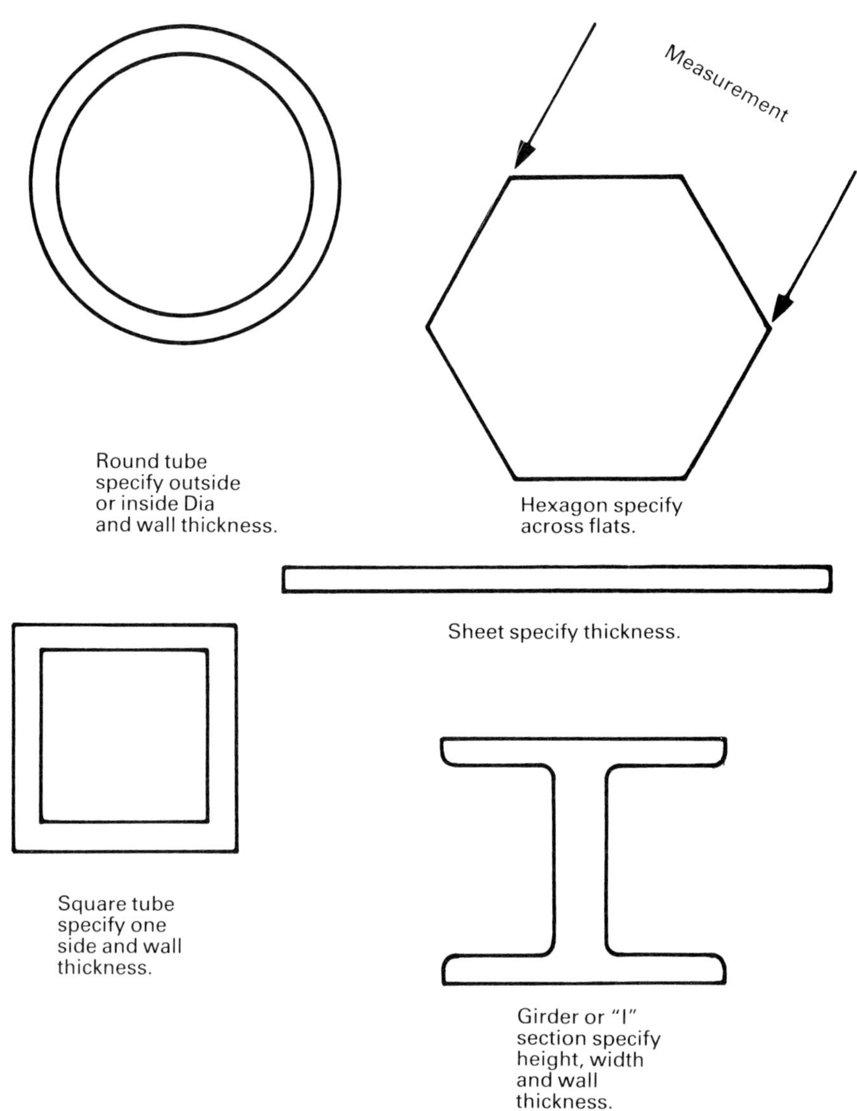
Measurement
Round tube
specify outside
or inside Dia
and wall thickness.
Hexagon specify
across flats.
Sheet specify thickness.
Square tube
specify one
side and wall
thickness.
Girder or "I"
section specify
height, width
and wall
thickness.

SHEET METAL THICKNESSES

Sheet metal is sold usually in thicknesses known as Standard Wire Gauge (SWG). Sometimes it can be purchased in metric measurements. There are no direct conversions but the following table may be of assistance.

SWG	Millimetres	Inches	SWG	Millimetres	Inches
8g	4.09	0.160	20g	0.91	0.036
9g	3.64	0.144	21g	0.813	0.032
10g	3.25	0.128	22g	0.707	0.028
11g	2.93	0.116	23g	0.607	0.024
12g	2.64	0.104	24g	0.560	0.022
13g	2.32	0.092	25g	0.511	0.020
14g	2.043	0.080	26g	0.460	0.018
15g	1.82	0.072	27g	0.420	0.016
16g	1.62	0.064	28g	0.374	0.014
17g	1.43	0.056	29g	0.345	0.013
18g	1.214	0.048	30g	0.315	0.012
19g	1.02	0.040			

PLASTICS

Some plastics may be used from time to time, in the main these will be for bearings or for heat insulating handles etc. The plastic for heat insulating handles comes in powder form. The tool to be treated is heated to a point where it is just hanging colour. The area to be coated is then plunged into the powder which sticks to the metal and melts into a smooth coating. For better results some form of blower is required and the hot metal suspended in a tin of powder being blown around. This gives a slightly more even coating. For bearings Teflon can be purchased in rod form. It can be machined as required and makes a reasonably easy running bearing which does not require lubrication. Nylon rod can be dealt with in the same way.

7 CUTTING METAL

One thing that you will certainly have to do during your model engineering is to cut metal, be it in the form of sheet or strip. Once again if you know what tools are available for each operation then you will be half way to success.

SHEARS

Tin snips are sufficient for cutting thin sheet metal. They come in two types, flat or curved nosed, the curved nosed ones being used mainly for cutting curves. They are used just like a pair of scissors and providing the metal is not too thick are quite useful. They do tend to make the metal curl up at the edges though and this can be a disadvantage where accuracy is concerned. An advance on tin snips are metal shears. These are either screwed to a bench or held in a vice. There is slightly less danger of curling up and heavier metal can be cut. They do take up valuable space in the workshop and are comparatively expensive for a limited amount of use.

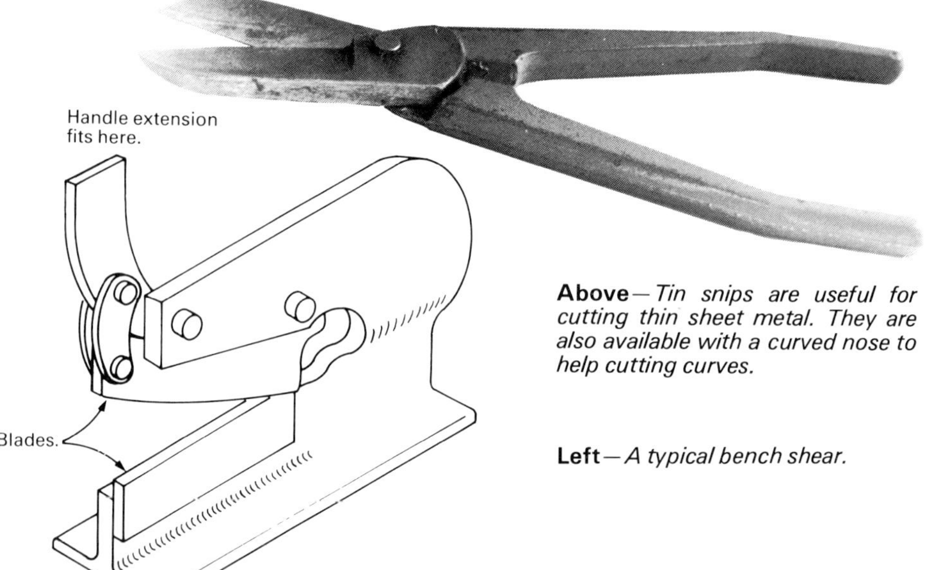

Above—*Tin snips are useful for cutting thin sheet metal. They are also available with a curved nose to help cutting curves.*

Left—*A typical bench shear.*

SAWS FOR SHEET METAL

A special type of hacksaw is made that is very good for cutting sheet metal of 20 gauge and above. As long as the metal has to be cut straight then the saw can be very accurate. Other saws available include some sold for dual purposes, but they are not in my opinion anywhere near as efficient as the hack saw type. For cutting shapes in sheet metal then, a coping saw or a piercing saw are the ideal tools. The piercing saw takes a somewhat finer blade than the coping saw, but it is surprising just how thick a sheet metal it will cut. In my opinion the use of a saw is far better than shears or snips as the metal is left in better condition.

Below, top to bottom—*A sheet metal saw, a coping saw, and a piercing saw.*

Below right—*A pad saw, useful for small work, and for using up broken hacksaw blades.*

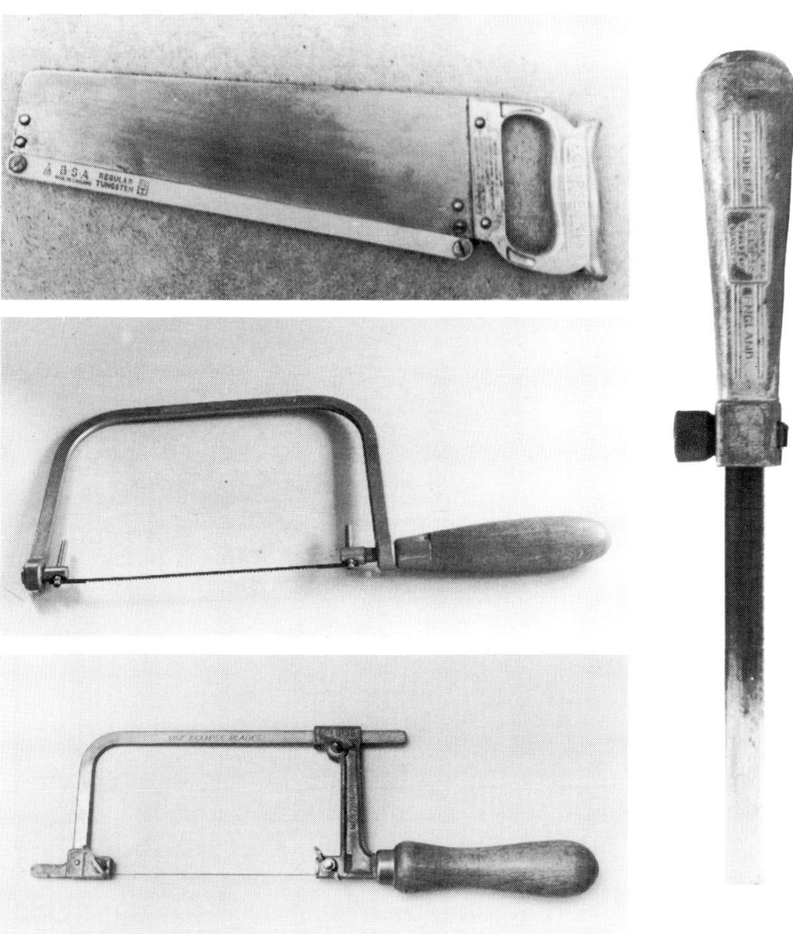

CUTTING METAL BARS

For bar and similar material you need a hacksaw. They come in two varieties, the type with a pistol grip and those with a straight handle. It is a matter of personal choice, but I do find that the straight handle type can be more delicately handled. Most hacksaws will accept blades of varying lengths and the blades can be purchased in several types. Standard blades are made of a flexible material and are designed to bend rather than break in use. This type do not last as long as the high speed steel blade which is much harder but will break much more easily. If you are a careless type of person a standard blade might be best for you. There are also dual types, allegedly high speed steel teeth and flexible backs to give the best of both worlds. Apart from varying lengths, the blades are sold with a varying number of teeth per inch. For the standard hacksaw there are four varieties 14, 18, 24 and 32. For power saws there are blades with a variety of numbers of teeth. It is fair to say that the finer the sawing you are to do the more teeth per inch are required. When putting a blade in the saw, the points of the teeth should point to the front of the saw. A strange anomaly arises here in that there is a school of thought that says when putting a blade in a coping or piercing saw the teeth should be pointing the other way. There seems to be no firm authority on this though, and I find that it will depend on the type of work being undertaken. As far as the hacksaw is concerned there is no doubt whatever and this also applies to a junior hacksaw, a smaller version of a hacksaw which is very useful.

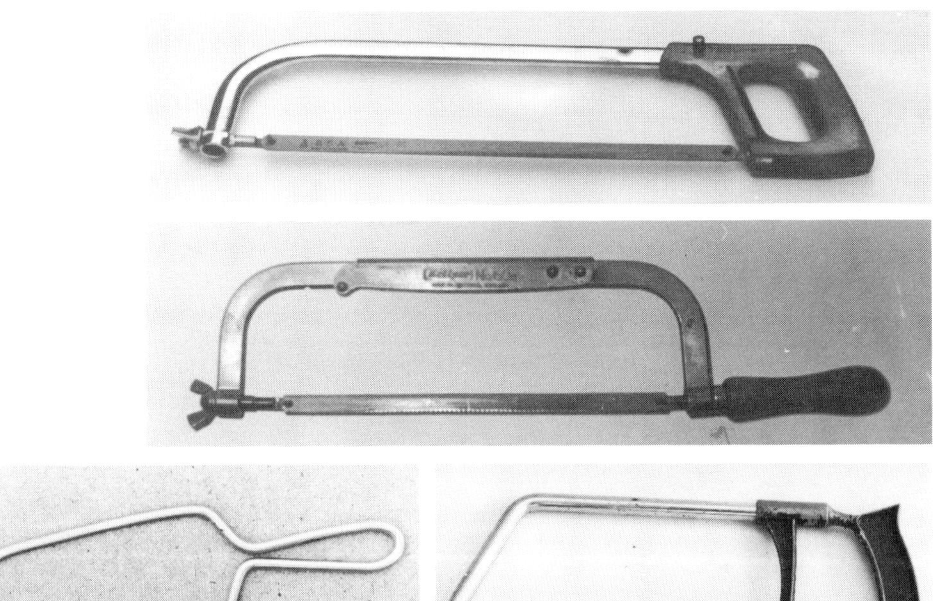

A hack saw fitted with two blades for cutting slots.

Opposite page, top—*A hacksaw with pistol type grip.*

Opposite page, middle—*A hacksaw with straight type of handle.*

Opposite page, bottom left—*A simple junior hacksaw, useful for small cutting work.*

Opposite page, bottom right—*A junior hacksaw of a more complicated type, with blade tensioning screw.*

USING SAWS

To use a hacksaw it should be gripped by the handle and supported at the other end. Remember there is only one cutting edge on the teeth and so it should be pushed into the work. If it is brought back whilst in contact with the work the only thing created is wear on the blade and so the rule must be to cut in one direction only. Make sure the saw is kept upright. Do not hurry, and keep checking how things are going—hacksaws appear to have minds of their own and will gladly cut at an angle when meant to be going straight. This is caused by hurrying the work, keep the saw in line and the trouble will not occur. A little cutting oil will also help. The saw should be pointing down at a slight angle in order to get the maximum work from each stroke. Take things particularly easily when you are nearly through the metal as there is a danger of the saw suddenly going through with a jerk and causing a nasty injury. Cut to the side of any scribed line where there is room for waste. A hacksaw does not just part the metal in two but actually removes a certain amount of metal and allowance must be made for this. Incidentally there is no reason why use should not be made of this fact and the saw used for cutting slots. If the slot it cuts is not wide enough then fit two blades next to each other and see if this does the trick.

Starting the saw cut off can be a problem. Sometimes there is a tendency for the saw to skid and mark the metal if you are not careful. Cuts can be started off by carefully pulling the saw over the work in the reverse direction, creating a score mark into which the blade will settle. Another idea is to scribe a second line some 2 mm or so away from the working one. This is called a witness mark and the idea is to keep the saw between the two. Because the strokes at the beginning are so light, if the saw does slip the witness mark will tend to prevent it going too far. It can also be a valuable guide for the main sawing. Very fine work can be started by putting the finger in an upright position and touching the blade. By making sure the blade is kept in contact with the finger nail for the first half dozen or so light strokes the line will start very accurately. A

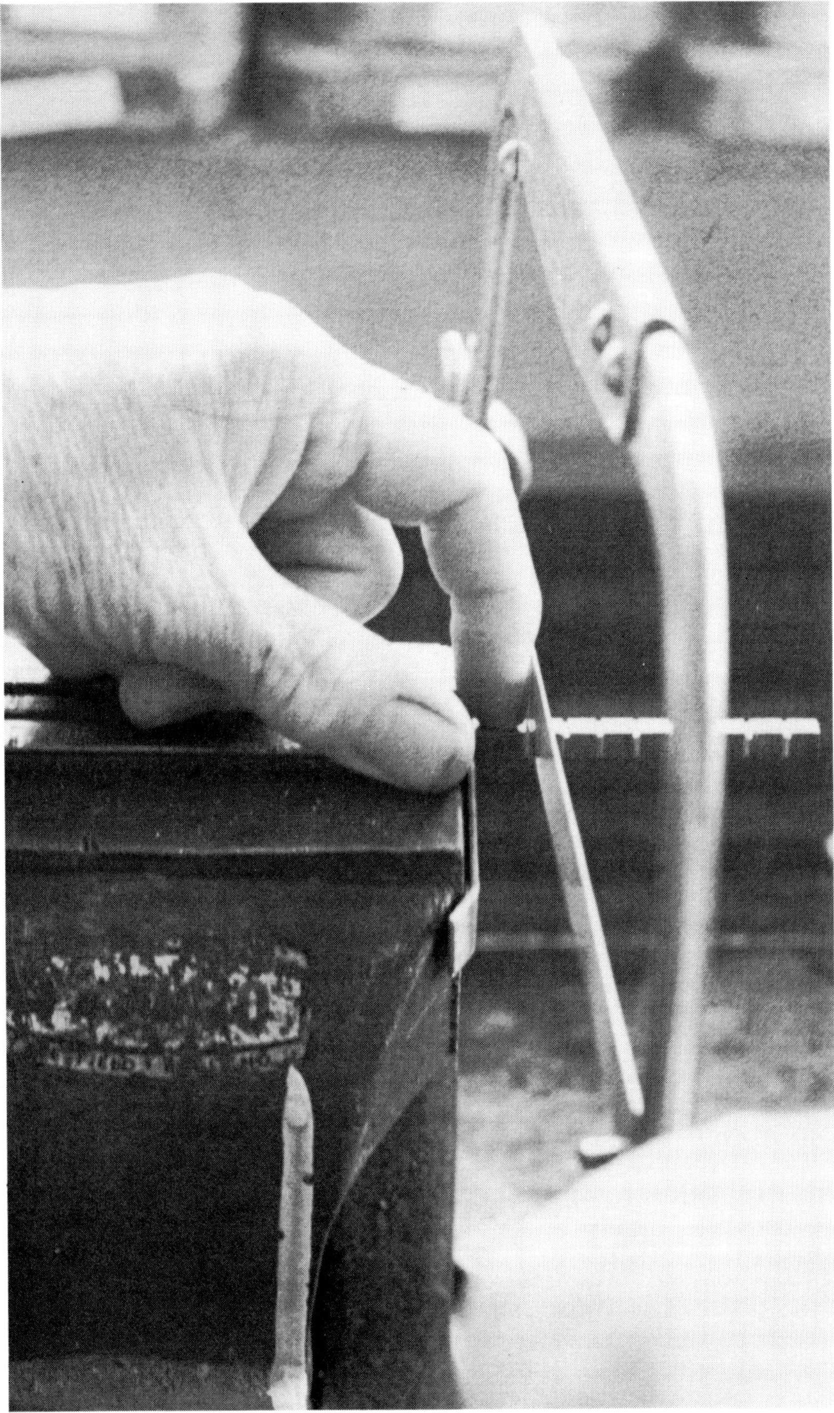

Above—*A hacksaw fitted with an abrafile blade. This device cuts round or straight and is useful for cutting out odd shaped holes. Its use is referred to in Chapter 3.*

Opposite—*Using a finger to guide the hack saw blade at the start of a cut.*

Below—*A hacksaw with blade fitted sideways to cut long material.*

Above— *The wrong way to cut flat material; not only is the metal being cut in the wrong direction but the saw is not tilted downwards as it should be.*

Below— *The right way to cut flat material is across the width.*

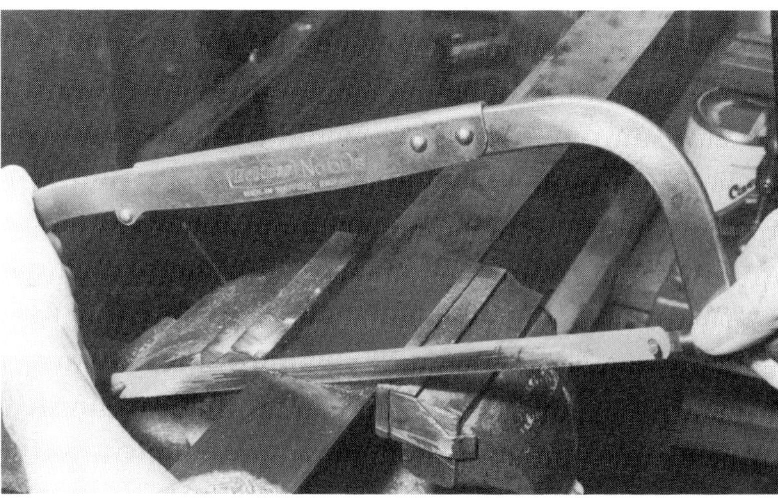

final little tip is to turn the saw blade sideways if the length of work is too long for the saw to go through in the normal manner.

When cutting square bar the saw should be used only from the position in which it started. Do not be tempted to keep sawing a bit from each side—this does not work. Rectangular bar and sheet should always be cut across the largest area, for the greatest accuracy. Again do not be tempted to turn the work round and cut another face, keep going until the job is completed. Do not be tempted either to cut through across the width, ie with the saw entering the smallest edge. This will almost certainly lead to lack of accuracy. When it comes to round stock, the same rules basically apply. In the case of large diameters however it is possible to turn the work round and keep making fresh starts, providing the marking out of the cutting line has been correct in the first place, and here again is a good reason for a witness line. Doing this saves getting the largest area of cut into a smaller one, which can cause the saw to wander. Also because the bar being sawn is round, it is easier to keep the sawing at each section accurate. A piece of masking tape wrapped round the metal will act as a guide.

Cutting a 4 in diameter bar. A piece of tape is used as a guide and the bar is cut round the circumference, a little at a time, to maintain accuracy.

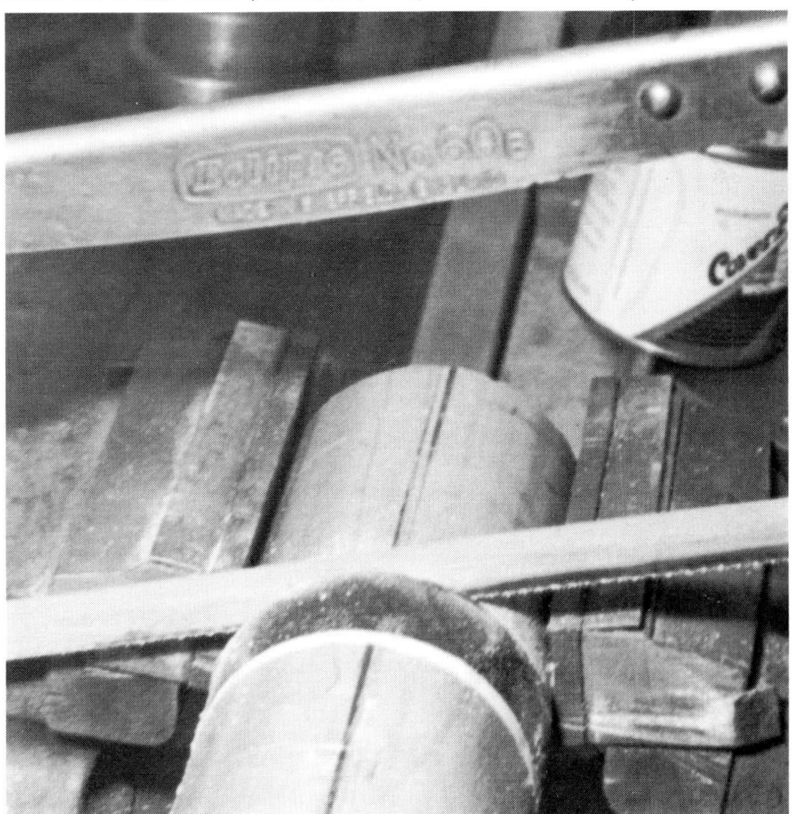

POWER CUTTING

Before closing the chapter a little thought should be given to mechanical means of cutting metal. Power guillotines for sheet metal can be purchased second hand at reasonable prices but whether or not they are likely to repay the money spent on them is a matter for speculation. They are also somewhat too large for most modeller's workshops. A little device on the market that will cut sheet metal very accurately indeed is called a nibbler and as it is operated from a DIY electric drill it requires no effort. It works by making a series of punched holes, so close together that they finish up as a straight edge. It will rapidly cut curves or straight shapes, but is limited to about 18 gauge brass and 20 gauge steel. Nevertheless I have had one for a long time now and I find it most useful.

Power saws are available, and for cutting bar material can be quite useful. They are somewhat bulky, but there are a couple on the market designed to fit on the bench. Many modellers use bandsaws and these are good because they will cut shapes as well as straight lines. There are two types; the two wheel variety, which means that there are two wheels for the blades to run on, and the three wheel version. The two wheel type is more accurate but also more bulky. It also suffers against the three wheel one in that the work which will go through is of necessity smaller. Although the three wheel type is noted for its lack of accuracy there are thousands in use, and I know of one that is used daily for cutting 12 mm thick steel plate. More recent in its entry into the market is the band saw that can also be used horizontally as a power hack saw. These are quite useful for cutting bar but I am told that, because of the blade width, they will only cut curves to a limited radius.

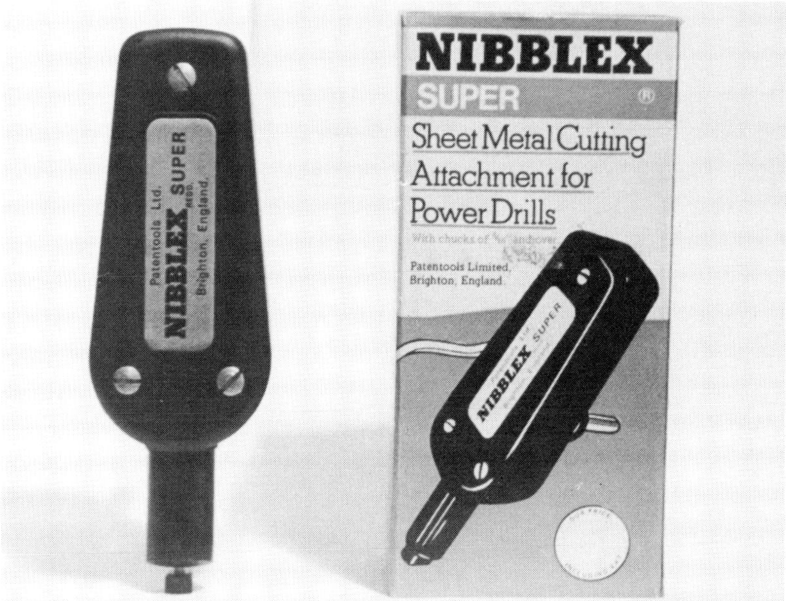

Above—*A Nibbler. It fits in an electric hand drill and is useful for sheet metal cutting.*

Below left—*A large power guillotine.*

Below—*A band saw, probably the most popular type of power saw for the home workshop.*

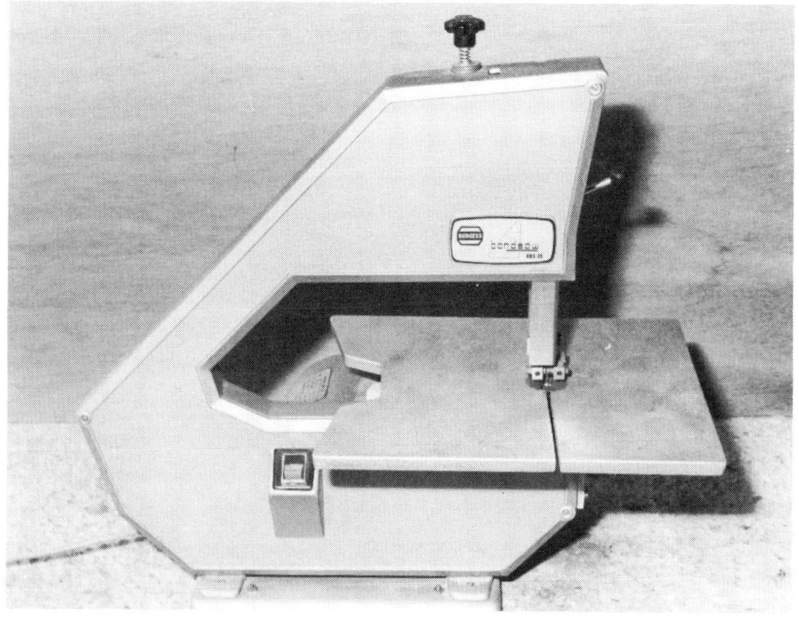

8 *FILING*

At one time filing was a skill which was insisted upon when apprentices were trained, and one which had to be perfected to a very high standard. Nowadays with the advent of milling machines and surface grinders it is rather a dying art. Even in the home workshop many tasks that were once done with a file are now put on a milling machine. This is a great pity as it is possible to get a lot of satisfaction from making something where high quality filing has played a part. It is probably fair to say that one reason for its decline in the workshop at home is that people do not really understand it. It can often take longer to set up a milling machine than it would to file a piece of metal to size.

First of all let us as usual think about the tools that are involved. Everyone knows what a file is, but how many know what grades are available? There are five, although sometimes it is quite difficult to get hold of two of the grades. The grades are known as rough, bastard, second cut, smooth, and dead smooth. Of these the second cut and the dead smooth are frequently not stocked by tool shops, unless they happen to be stockists of a very high standard. Dead smooth files are often called Swiss pattern precision files and they can be obtained in various subgrades. Different manufacturers give numbers in different ranges and so it is as well to check with the supplier when buying in order to know

Files are available in two types of cut as well as different grades of cut. The single cut has teeth in one direction only, the double cut has the teeth in a diamond pattern. Although, more often than not, the finer grades are single and the coarser are double cut there is considerable overlapping. The single cut gives a better finish when draw filing but is less suitable for cross filing large surfaces.

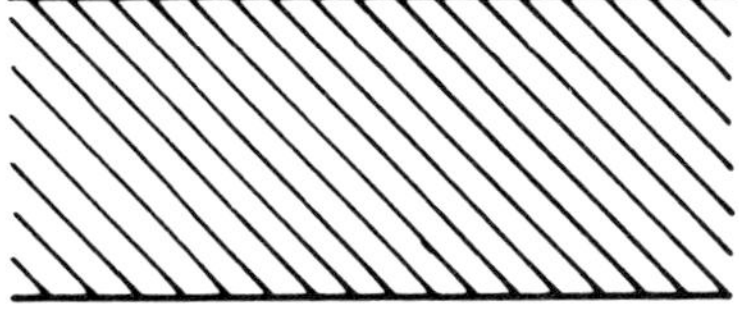

Single cut

Double cut

what you are getting. Also available are two cuts for really heavy work. They are known as dreadnought and millenicut and frequently can be bought as a single tool with one of these cuts on each side. They have a quite remarkable ability to remove metal and are, to a degree, non-clogging.

Apart from the cut you will also need to know the shapes and lengths that are available. Lengths will vary from 4 in to 14 in, the work that you are going to use the file for governing the length required. If you are trying to remove a great deal of metal then a 4 in file will be of little use, equally a 14 in file will hardly be appropriate for filing the teeth on a clock gear.

For removing large amounts of metal two special cuts are available, dreadnought above is the coarsest and millenicut below is just a little smoother.

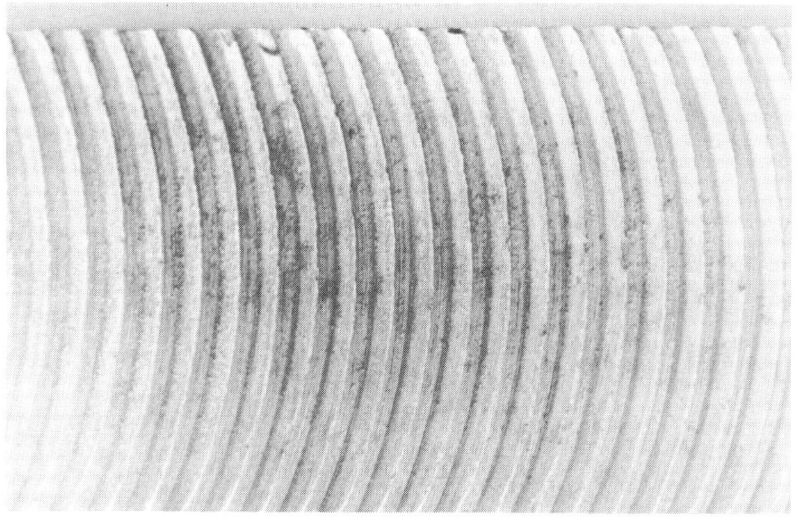

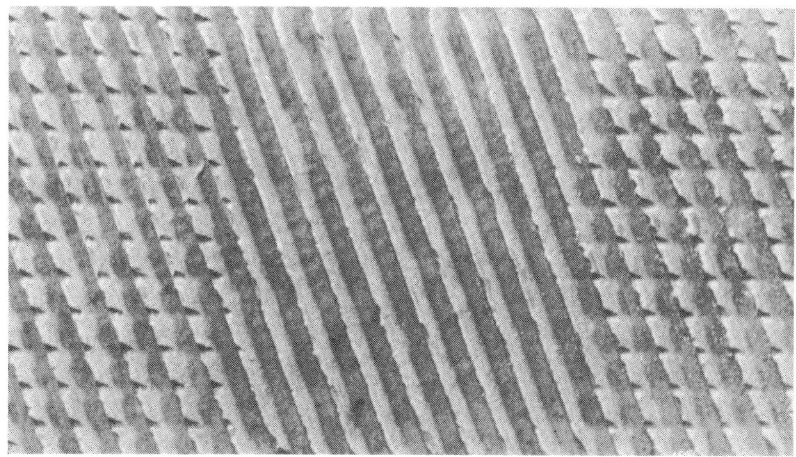

Files come in various shapes, each with its own use.

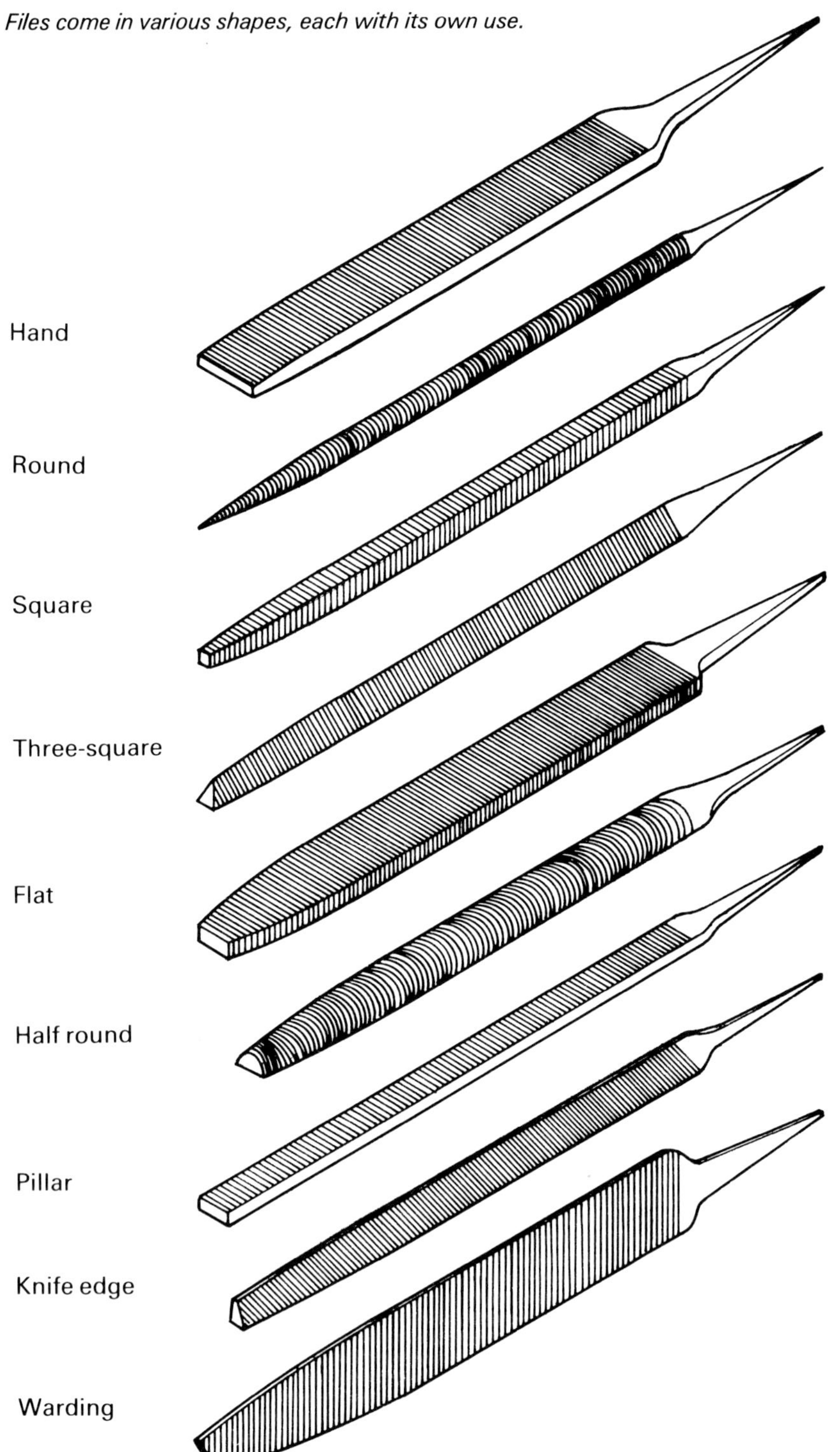

There are nine shapes of file available. Hand, which is the normal type, is parallel in width, but with a slight tapering of thickness towards the end. Flat is similar to hand but the end tapers in both thickness and width. Pillar is again similar to a hand file but much narrower. Warding is like a flat file but very thin, usually so thin as to be almost flexible. A half round file, as the name implies, is half round in shape, with a slight taper towards the end. Square files are of equal width on all four sides, tapering towards the end. Round (sometimes called a rat tail) as one would expect it is round in section and again tapers towards the end. Sometimes the teeth on a round file spiral from front to back. Three square is triangular in shape, with a tapering end. Generally files are purchased without handles—they have a tang, or pointed end, which allows them to be driven into a wooden handle. Under no circumstances whatever should a file ever be used without a handle—it is extremely dangerous to do so. Wooden handles that do not fit properly or are split should be replaced. Some manufacturers are now supplying the more common types of file with a moulded plastic handle ready fitted. These are excellent and are to be highly recommended. Swiss type files come in slightly different shapes, but the differences are so slight that they need make no difference when selecting the file requested for the purpose. For small work needle files are available. Usually these can be used as they are without the need to fit handles. They consist of very small files in a variety of shapes, which are often sold in sets. There are no rough ones but a number of different cuts are available, usually varying with the individual manufacturer. Many have smoother teeth than one would normally expect to find when selecting the required tool and it is worth checking before buying as some of the very smooth ones will do little more than polish metal.

Files are precision tools and should be treated as such. Under no circumstances should they be put in a heap in a tool box or drawer as rubbing against each other will take the edge off them. They should be kept separately in racks. Needle files can be kept in wooden blocks. All files will need to be cleaned regularly, this can be done by rubbing a wire brush over them. Small slivers of steel will embed themselves in the files and these will not only cause the file to blunt, but will badly scratch work. The reason for this is that the friction hardens the little pieces until they become harder than the file itself. A small piece of brass sheet rubbed across the teeth will remove them. Files that have been used on steel will not cut brass. (The same applies to hack saw blades). Once a file has been used on steel I paint the handle so that I know that I cannot expect to use it when filing brass. Files that have blunted can be sharpened a little by standing them in a solution of Hydrochloric Acid at about a ten per cent strength. Leave them overnight and then wash in water. The treatment is quite effective.

Most filing is done by rubbing the file across the work. Hold the handle in one hand and the end of the file in the other. Push down and across away from the body. Lift the file off the work and repeat the operation—do not keep the file in contact with the work on the return. Try and keep the file perfectly level, with a little practice it is surprising how easily it becomes natural to do this. Do not hurry and check frequently to make sure that things are going right. There is little point in filing away rapidly and merrily only to discover that for quite a while the work has been out of square or something similar. Before filing it is a good idea to rub chalk in the teeth or put on a little cutting oil. This saves the teeth getting clogged up. Even so the file will need frequent wire brushing to keep it clean.

Filing grooves, whether they be round, square or vee shaped can be somewhat awkward. I find it best to start with a faint hacksaw cut in the required place, then to put in a small vee and finally to take out the shape I require. Trying to file direct into the shape, more often than not, results in the file skidding across the work and leaving nasty scratch marks. When filing vees and squares make sure that the file does not twist sideways in your hand, this is a common cause of things going wrong. To ensure a neat straight edge on filing of any type, an old hacksaw blade can be put beside the work in the vice. It is then possible to file right down to it and know that the edge will be straight. To make a curve the file should be rotated as it is passed over the metal. Large inside radii can also be made in this way, with either a round or half round file, using a draw filing action.

Draw filing is a method employed to obtain a fine finish on a piece of metal. It follows, therefore, that it is carried out with a fine file. It is the exact opposite of normal filing in that the file is drawn along the metal rather than taken across it and that the file is kept in contact with the

For normal work the file should be pushed across the work, lifted off and then pushed across again. When draw filing, the file is slid lengthways along the work.

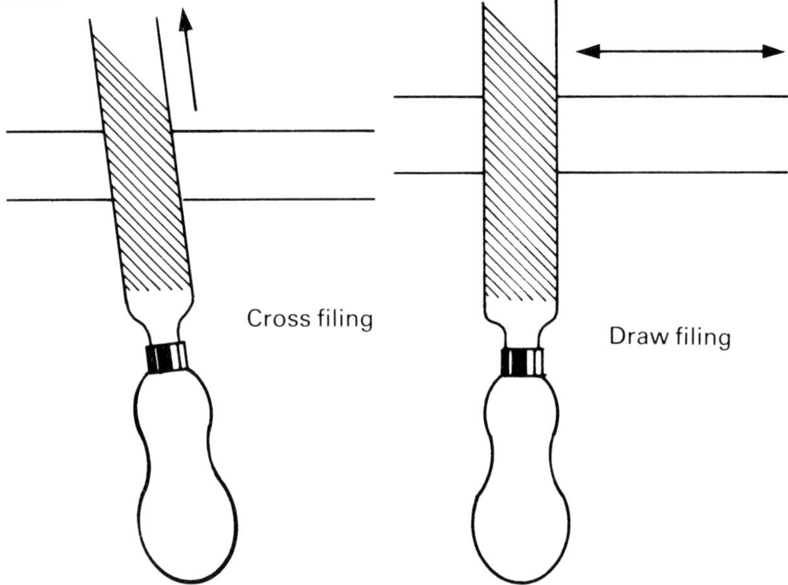

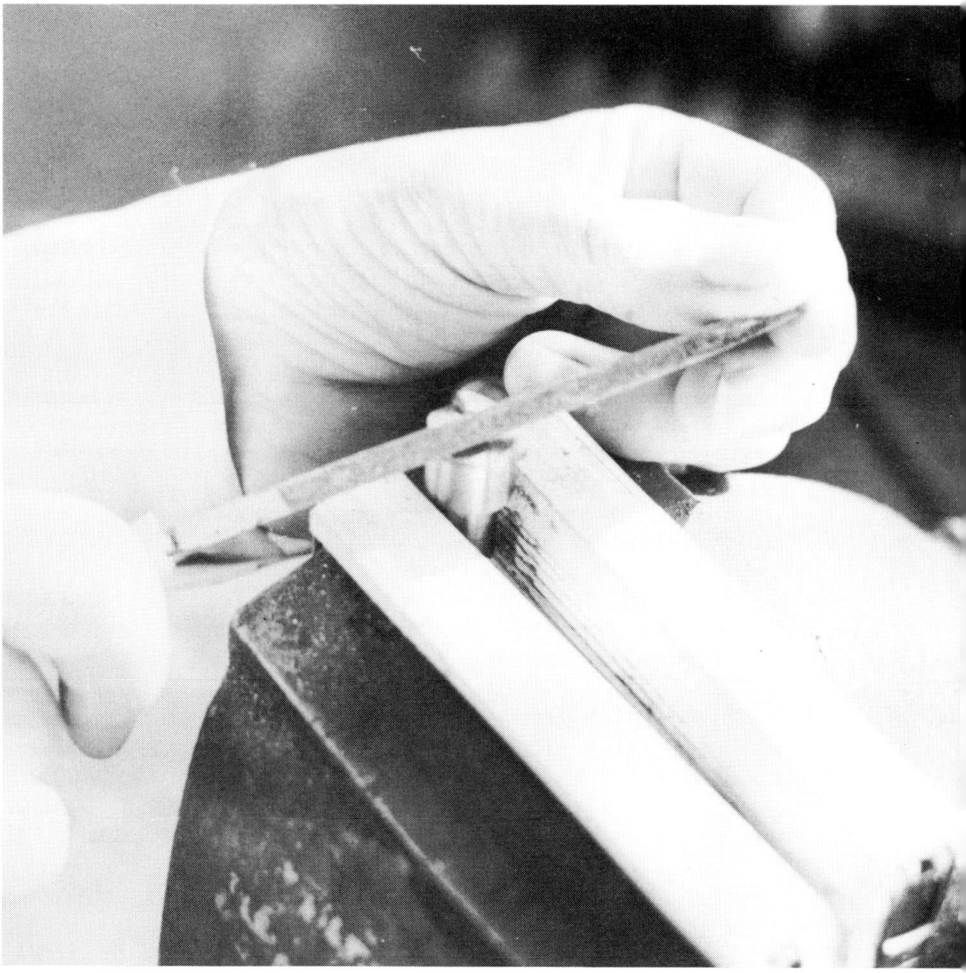

The file should always be supported at the end by the hand.

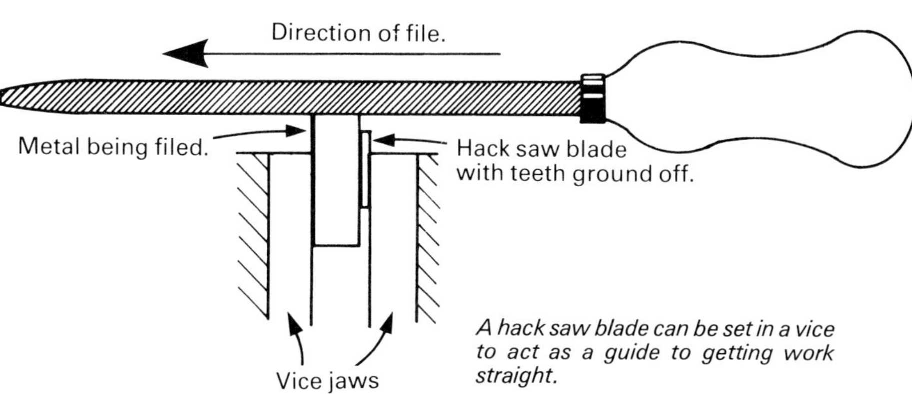

A hack saw blade can be set in a vice to act as a guide to getting work straight.

metal on the return. Again chalk will help prevent clogging and the method can result in a particularly professional looking finished piece of work. Again there is the problem of the possibility of the file being pushed down at the ends of the work and creating a bowed effect and care must be taken to avoid this happening. The operation can be carried out on flat surfaces of work and frequently in such circumstances it will not be possible to retain the work in a vice. The answer is to hold the work on a piece of wood with panel pins, which is quite sufficient to prevent it moving, and if a piece of wood slightly longer than the work is used it can help prevent the file from rocking. In the case of a small piece of work it is sometimes a good idea to put the file in the vice and rub the work along it. In this way there can be no question of the finished piece not being flat.

Although a fine finish can be obtained by draw filing, even better results can be obtained from the correct use of abrasive papers or cloth. At one time this was limited to emery cloth, now there are several varieties of material available which work equally well if not better. Carbide paper is one and this can be distinguished by its brown colour. Another is known as 'wet and dry' paper and is mainly sold in car accessory shops. It can, as its name implies, be used wet or dry. Either way it gives a superb finish, and if wetted with a little oil before the final work can also help to prevent rust. Whatever the paper or cloth it should either be folded round a file or glued to wood to use. Emery sticks, which consist of emery paper glued on wood, can be purchased and they are excellent. Just recently a plastic strip has appeared on the market to which such abrasives can be glued—they do a good job as well. Use of any abrasive paper or cloth is simply a case of rubbing on the metal until satisfied that the finish suits you. The paper or cloth should be rubbed in one direction only, lengthwise usually gives the best results.

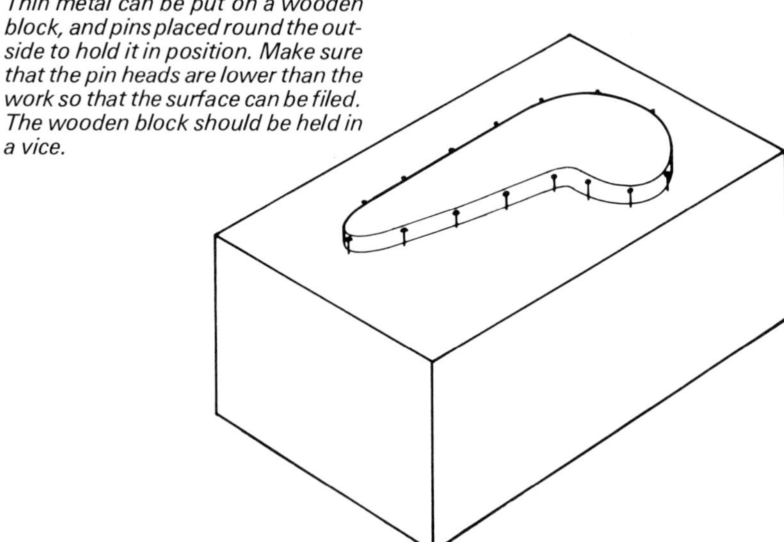

Thin metal can be put on a wooden block, and pins placed round the outside to hold it in position. Make sure that the pin heads are lower than the work so that the surface can be filed. The wooden block should be held in a vice.

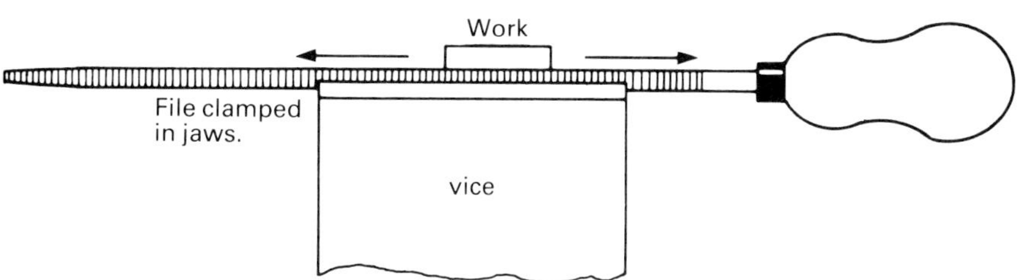

For some draw filing it is better to place the file in the vice and rub the work on it.

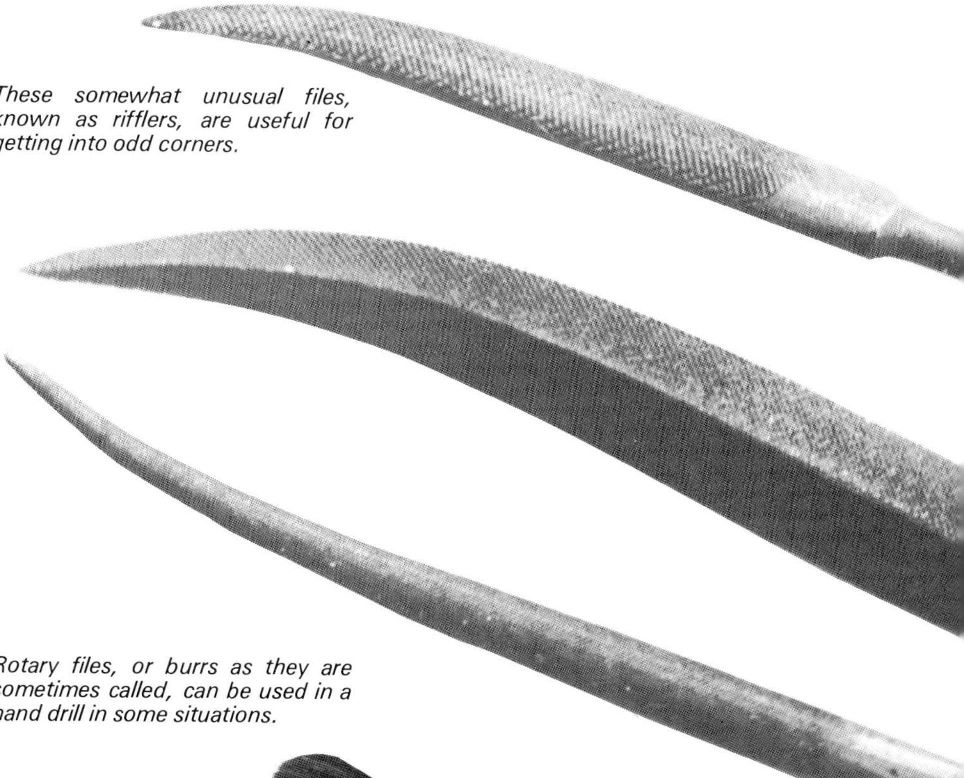

These somewhat unusual files, known as rifflers, are useful for getting into odd corners.

Rotary files, or burrs as they are sometimes called, can be used in a hand drill in some situations.

9 *DRILLING*

Whatever you make in the workshop will probably need to be drilled. It is highly probable that at some time or other holes will be required of all sorts of sizes. I doubt if there is likely to be a reader of this book that will not know what a drill looks like, but I wonder how many realise just how many different types of drills there are. Incidentally in this context I am referring to the bit that does the actual hole making, rather than the appliance in which it is held. This can be somewhat confusing as both are referred to as a drill, but the part I am talking about is often also called the bit.

To use the bit we need some sort of appliance. A simple hand drill known as a wheel brace can be useful, particularly where extreme accuracy is not a necessity. There is no great secret in using one of these, other than to apply a medium sort of pressure at the top and to keep the wheel turning, rather than keep starting and stopping. A good sharp bit is essential. Most people these days are equally familiar with the DIY type of electric drill. These will go through quite thick metal and are very useful. With a vertical stand to support them they can be quite accurate. I know of one model engineer who uses such a drill in a stand all the time and has no problems. They come in a variety of sizes, with chucks up to ½ in capacity. Some have variable speed which is very useful as the larger the drill size, the slower the drilling machine speed. The third type of device to be considered is the drilling machine. This is a purpose built machine with a solid column supporting the drilling head and the table on which the work is laid. They come as either bench or floor standing models. My own preference is for the bench version as I can then store things underneath. Although I have a very old floor standing one as well, which was a throw out from a factory, I rarely use the extra height it gives me.

Drilling machines are available in all sorts of sizes and it largely depends on what sort of work it will have to do as to what size you will require. If the work is to be small, then one of the little precision machines will be ideal. The chuck capacity of these is usually no more than 6 mm, but for small work this is sufficient. They are usually very

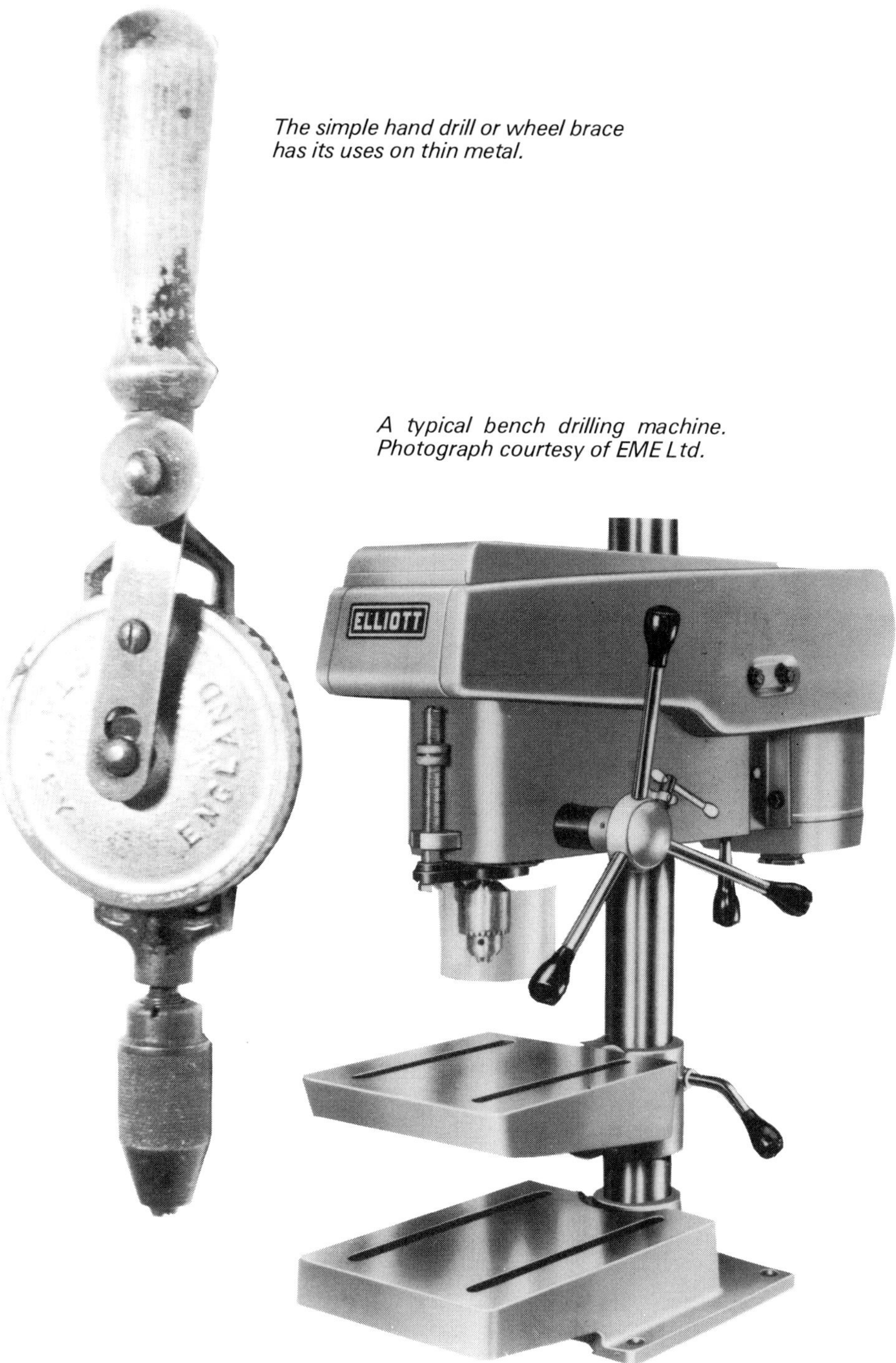

The simple hand drill or wheel brace has its uses on thin metal.

A typical bench drilling machine. Photograph courtesy of EME Ltd.

A small precision drilling machine with a capacity of 6 mm. It is shown here being tested for accuracy.

accurate and have a very good range of speeds. When it comes to the larger models there are several variations in design. Some models allow the table on which the work is stood to be tilted at an angle or rotated, refinements which can be of use in some operations. Another useful extra often available is a rack to raise and lower a table. With the work on it, the drilling table can be quite heavy and the ordinary clamp, when undone, leaves the operator holding all the weight. With a rack at the side the table can be wound up and down irrespective of anything that might be on it. A variety of speeds is important too—the variety available to you being largely dependant on how much you are prepared to pay. Do get as big a range as you can afford. Chuck sizes vary and sometimes the same machine is available with either a chuck of 8 mm capacity or one of 12 mm. If the larger chuck is specified there should also be a larger motor. I cannot believe that many manufacturers are likely to put an extra large motor on the machine with the smaller chuck, so what happens is whilst the machine is made for the smaller size, a bigger chuck is fitted. This is not good practice and such models are best avoided.

WORK HOLDING

Where possible work being drilled should always be clamped firmly to the table of the drilling machine. If this is not possible then it should be held in as large a machine vice as possible. This will prevent it riding up the bit and flying round. If possible then, clamping a machine vice to the table and putting the work in that is the ideal solution. A machine vice is a vice with a flat base which enables it to be laid flat on a machine or a bench. They come in a variety of sizes. Some of the more expensive ones have swivel bases so that they can be rotated after being clamped to the table. I use one that also tilts, useful for drilling holes at an angle. It is graduated in degrees and saves me having to tilt the drilling machine table. This has meant that I have been able to set the table with great care and I am not going to have to move it, with the inevitable loss of accuracy that would result.

Right—*A tilting and rotating machine vice that can help certain operations.*

Below—*A typical small machine vice suitable for drilling work in the home workshop.*

DRILL BITS

Drill bits are made in a variety of types, diameters and lengths. As far as diameters are concerned there is a complete range of metric drill bits available increasing in size at 0.1 mm steps with intermediate steps of 0.05 at certain sizes. In imperial measurements the range of drill bits increases in size at stages of 1/64 in. There are also letter and number sizes. These were sets of drills which used to be obtained in numbers from 0-80 and letters A-Z. They were odd fractional sizes, many produced to match with threads that were then in use. You will still find them referred to, but they have now been superseeded by the metric series. The type of drill for which you will have most use is called the jobbers drill. It has a straight shank and is available in carbon steel or high speed steel, although it is hardly worth considering the carbon ones when working in metal. (They do have their uses though when working in wood, if expense is to be spared.) Drill bits are also available with taper shanks, to fit direct into morse tapers. Many drilling machines are fitted with morse tapers for holding the chucks. It is also possible to purchase drills with stepped shanks, so that a larger drill than the chuck will accommodate can be used. These stepped drills are sometimes called blacksmith's drills. Drill bits may be purchased individually or in certain sizes as sets. They should always be stored in racks and not allowed to rub against each other in drawers.

Drill bits consist of a piece of metal twisted, or more likely machined, into a spiral. There is a point on the end, and this is ground to a precise angle. The two twisted sections going along the length of the drill are referred to as the helix, and should again have the correct angle for the type of drill. The piece up the centre is known as the webb and on standard jobbers drills this starts as a thin section at the point and increases in thickness along the length of the drill. Different types of drills are available for different metals, but most model engineers tend to make do with the standard drill for everything.

If the drill chuck is not large enough it is possible to use a stepped drill. This does place extra strain on the bearings of the machine though, and is not to be encouraged as a regular practice.

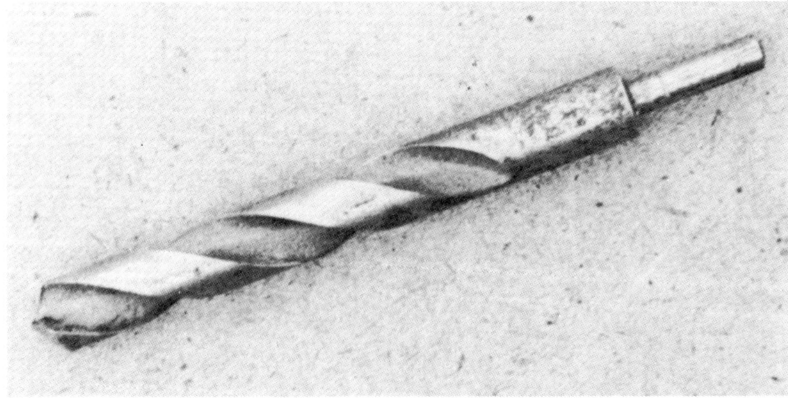

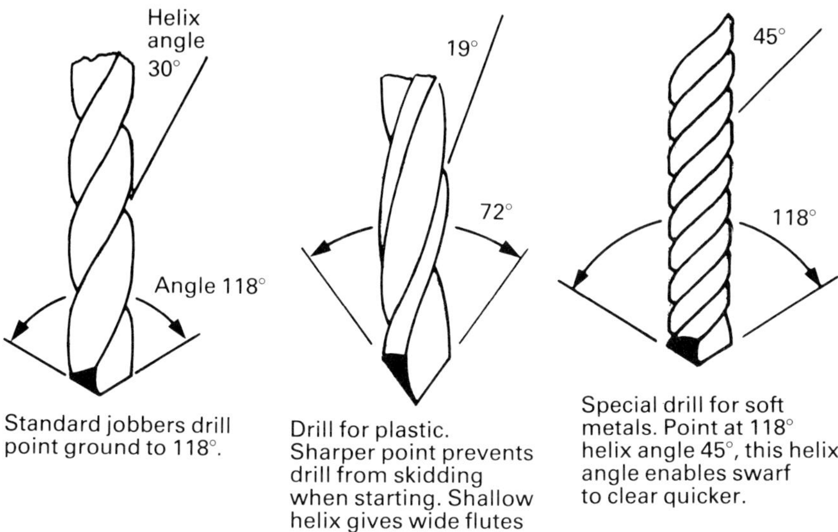

Helix
angle
30°
Angle 118°
19°
72°
45°
118°
Standard jobbers drill
point ground to 118°.
Drill for plastic.
Sharper point prevents
drill from skidding
when starting. Shallow
helix gives wide flutes
prevents build up of
heat.
Special drill for soft
metals. Point at 118°
helix angle 45°, this helix
angle enables swarf
to clear quicker.

Different types of drill are made for differing work. It is probably fair to say however that unless a great deal of work is to be done on a particular type of material, the jobbers drill will cope.

SHARPENING

When a drill is sharpened it is essential that both sides of the point are not only ground to the correct angle but also that they are perfectly even. If they are ground in such a way that the two sections of the point are different lengths, then the drill will cut oversize. It will also impose extra strain on the drill bit and may result in a breakage. As a drill is sharpened regularly it will obviously become shorter in length. This will result in the webb becoming much thicker than required, and this will have to be thinned by grinding with a fine grinding wheel. This is not possible with smaller size drills, so when the webb thickens it is as well to throw them away and start again. This applies to broken drills too, it may be possible

Right—*A typical small grinder for use in the home workshop.*

Below—*A correctly ground jobbers drill.*

to grind the drill to a point again, but the thickened webb will have to be dealt with. Common faults in drill grinding are: point angles too sharp or too flat; uneven angles; one side of the point longer than the other; insufficient clearance angles on the points. The sketches should show what is meant by these faults. They can largely be avoided by use of one of the special grinding jigs available on the market.

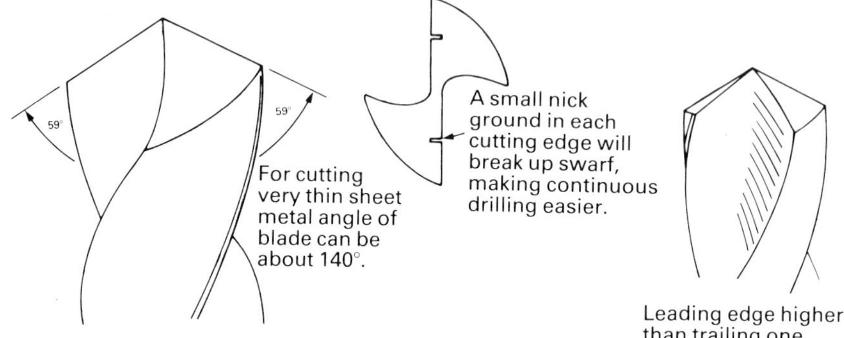

Correct cutting angles for jobbers drills.

DRILLING HOLES

Before actually drilling the hole the place where it is to be situated must be marked out with a centre punch. The hole should be started with a centre drill, and then the drill bit run through afterwards. The drill should be run at as near the correct speed as possible, and a suitable cutting fluid used. Do not put too much pressure on the work, but equally do not go too lightly. The first will cause the drill to bind and the second to rub. It is possible with a little experience to feel the drill cutting its way through the metal. To get the hole really accurate it should be boxed. This means scribing a small square exactly evenly round the centre punch mark. Run the point of the drill or centre drill lightly into this making the barest identation. It is now possible to see whether the hole will be in the correct place by comparing it with the square. If it is not right then move the work over in the right direction and very lightly run the tool in again. This should have the effect of pulling the indented centre mark over toward a central position. If not right try again, gently, until the correct position is found. When it is right drill through.

For accuracy holes should be boxed. This drawing shows how to do it.

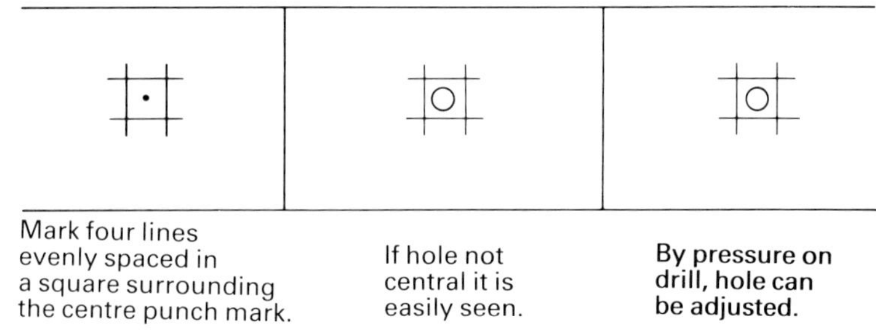

Mark four lines evenly spaced in a square surrounding the centre punch mark.

If hole not central it is easily seen.

By pressure on drill, hole can be adjusted.

LARGE HOLES

There is a limit to the size of holes that can be drilled with jobbers drills. If larger ones are required there are other tools that will do the job. A hole saw is a circular saw blade, mounted on a drill. The drill is run through the metal and the saw blade then starts to saw out the hole. These are quite effective, but are not as accurate as possibly one would like. Tank cutters are drills with a bar attached to them, a cutting tool through the bar. The drill is passed through the metal and the cutter then rotates, cutting the desired size of hole. They can be used very accurately with care, and are easily home made. A cone cutter is a tool in the shape of a cone as its name implies. A hole is first drilled with a normal drill, and then opened out with the cone cutter. It is effective and used with care can be highly accurate, the only slight problem being a very tiny taper on the finished hole edge. This can however have advantages where a fitment has to be soldered in as it makes a neat arrangement for the solder to settle into.

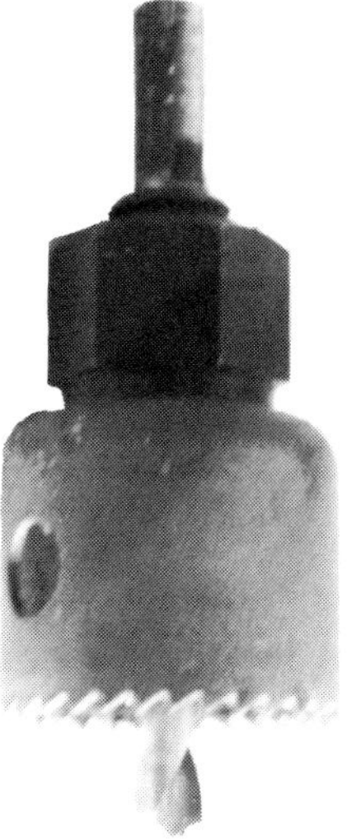

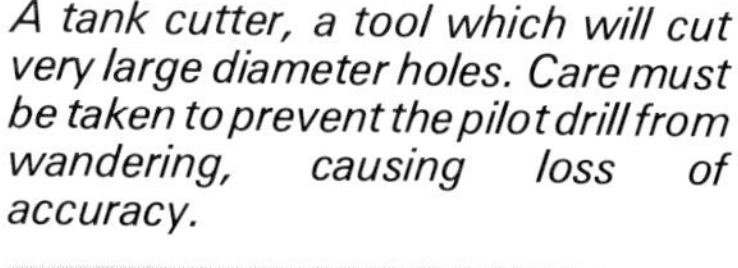

A tank cutter, a tool which will cut very large diameter holes. Care must be taken to prevent the pilot drill from wandering, causing loss of accuracy.

Above—*A hole saw, useful for large diameter holes but lacking a little in accuracy.*

Below—*A cone cutter will cut large holes very accurately, although with a slight taper.*

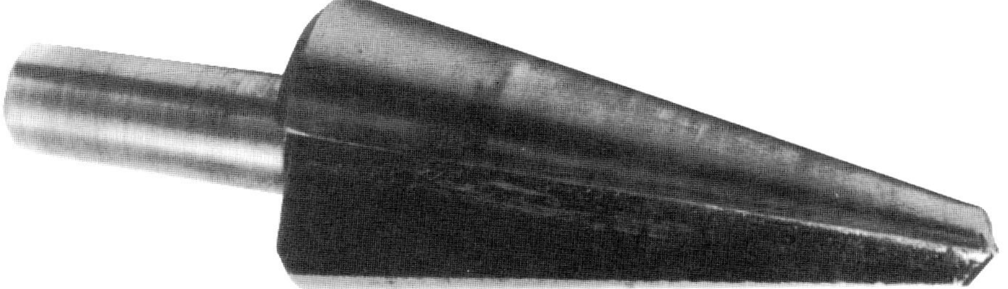

DRILLING THIN SHEET METAL

Drilling of thin metal is always a problem. A normal drill will make a hole of irregular shape. It is possible to make special drills that prevent this, or I have found that when used on thin brass the type of drill bit sold for cutting holes in wood with the household type drill will work very well. Hole saws can be used but problems arise with the pilot drill bit wandering before the saw has taken effect. Small punches work well, and the usual type consist of a punch and die with some system of screwing the punch into the die via the metal. Punches can be made at home, but these are usually made so that the punch is tapped through the metal with a hammer. Such devices are very useful for making thin washers.

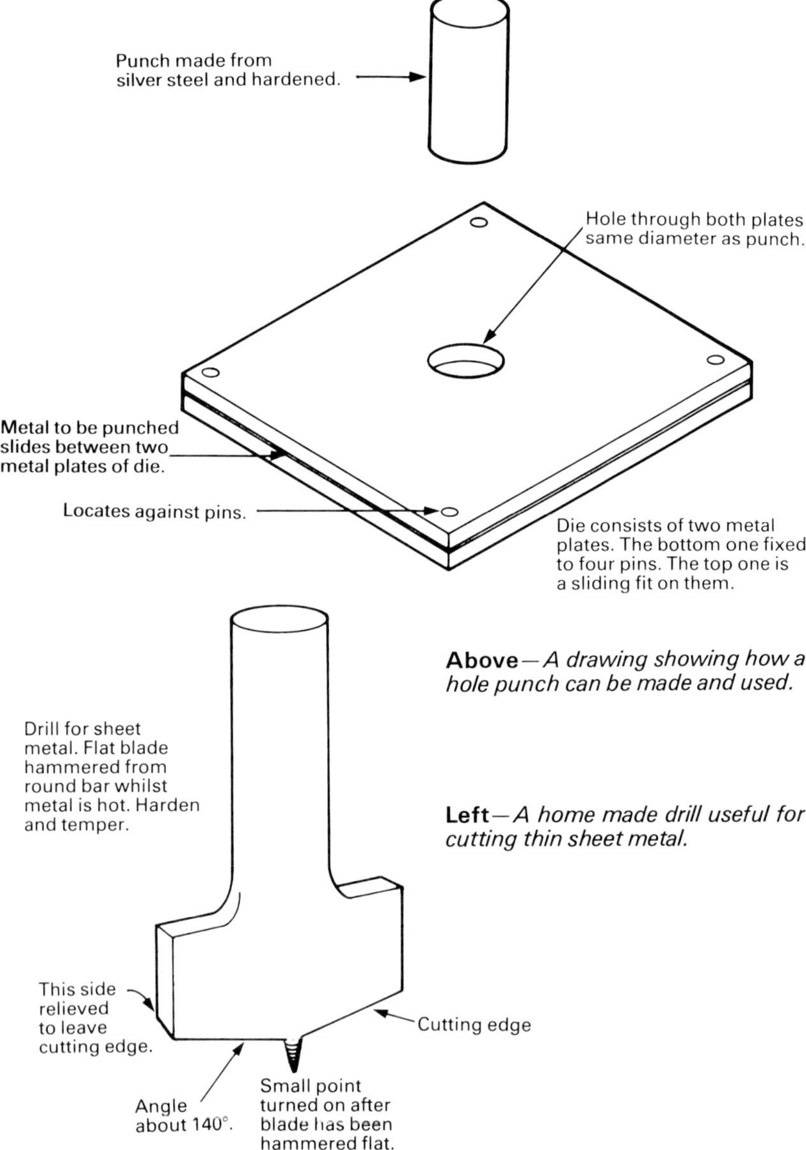

Punch made from silver steel and hardened.
Hole through both plates same diameter as punch.
Metal to be punched slides between two metal plates of die.
Locates against pins.
Die consists of two metal plates. The bottom one fixed to four pins. The top one is a sliding fit on them.
Drill for sheet metal. Flat blade hammered from round bar whilst metal is hot. Harden and temper.
This side relieved to leave cutting edge.
Cutting edge
Angle about 140°.
Small point turned on after blade has been hammered flat.

Above—*A drawing showing how a hole punch can be made and used.*

Left—*A home made drill useful for cutting thin sheet metal.*

SMALL DRILLS

For very small holes in the order of fractions of a millimetre we need a somewhat different approach. Very high speeds are needed and the feed of any drilling machine must be very light. Drill bits will need to be mounted in pin chucks before putting in the normal chucks, or in small pieces of brass rod, being held in with soft solder. In the case of drilling fine holes in very thin metal and plastic, it is possible to hold the drill in a pin chuck, or in its piece of brass rod and rotate it by hand, with an absolute minimum of pressure. Tiny light weight hand held drilling machines are also available and can be used effectively. These are usually battery operated.

Very small drill (this one is considerably less than half the diameter of a normal household pin) should be put in pin chucks or soldered to brass rod like this. The number 78 represents the size of drill, a range which is no longer available except from specialist stockists.

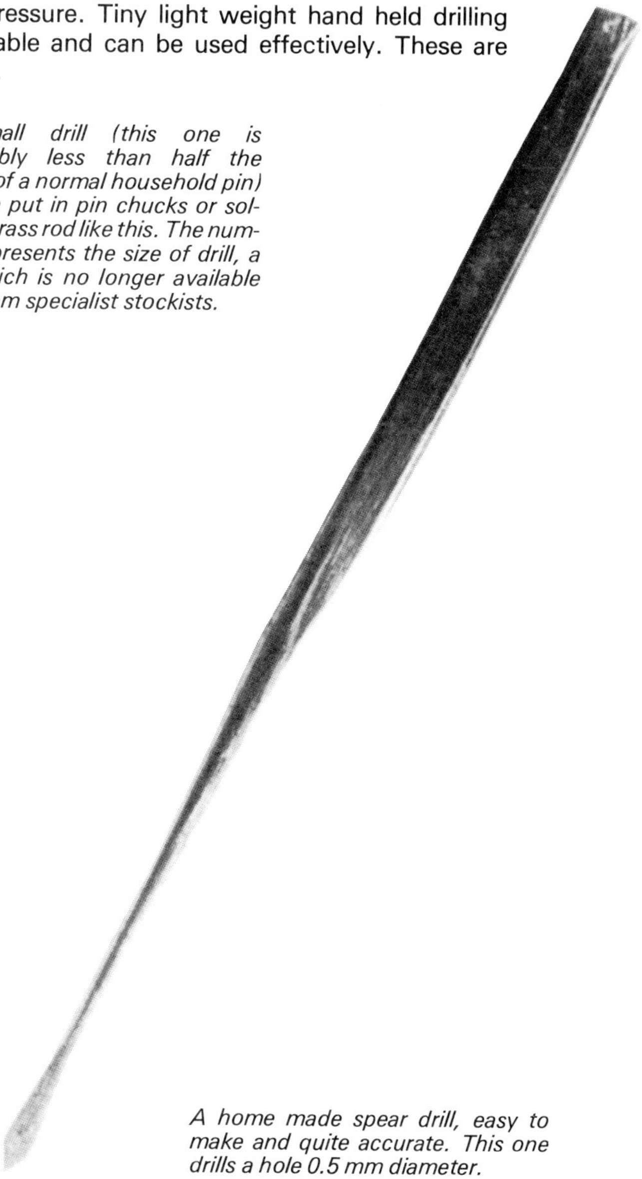

A home made spear drill, easy to make and quite accurate. This one drills a hole 0.5 mm diameter.

REAMERS

An ordinary twist type drill does not drill 100 per cent accurately. If you need to be very accurate then a reamer should be used. These are invariably made of high speed steel and are fluted. They are designed for use by hand or machine; the hand reamer is tapered towards the end and has straight flutes, whilst the flutes on a machine reamer travel in a slight spiral. The hole in which a reamer is used should be within a fraction of the size at which it is to be finished. If the work is to be done by hand the reamer is held in a tap wrench and rotated, making sure that the tool remains upright. When using a reamer in a drilling machine the work should be put in a machine vice but the vice left unsecured on the drilling table, just being held by hand. To obtain the maximum accuracy from a reamer it should be able to float, and if both the work and the reamer are securely held this is prevented.

Right—*A hand reamer. This one is for cutting tapered holes for taper pins.*

Below—*A machine type reamer.*

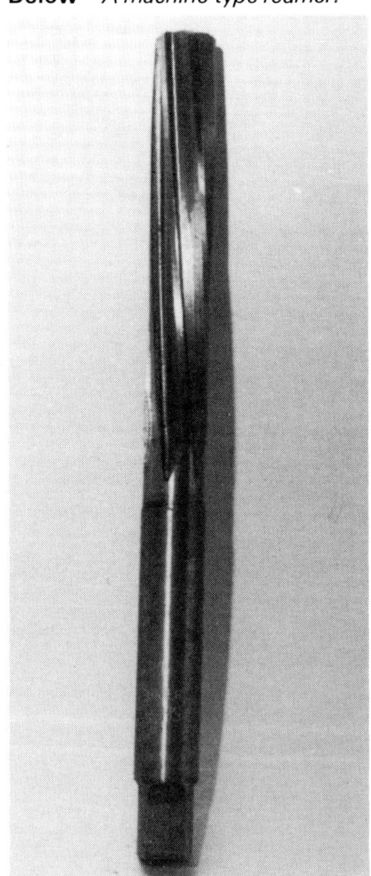

A countersink bit.

TABLE OF APPROXIMATE DRILLING SPEEDS

Drill Diameter	Type of metal Cast iron	Steel	Aluminium	Brass
3 mm	2550	1600	9500	8000
4 mm	1900	1200	7200	6000
5 mm	1530	955	5700	4800
6 mm	1270	800	4800	4000
7 mm	1090	680	4100	3400
8 mm	900	600	3600	3000
9 mm	850	530	3200	2650
10 mm	765	480	2860	2400
11 mm	700	436	2600	2170
12 mm	640	400	2400	2000

Speeds in revolutions per minute

BRITISH STANDARD PIPE THREADS TAPPING SIZES

Thread size (diameter of pipe)	Tapping size	Clearance size
⅛	8.75 mm	9.80 mm
¼	11.80 mm	14.00 mm
⅜	15.25 mm	18.00 mm
½	19.50 mm	21.50 mm

MODEL ENGINEER THREAD TAPPING SIZES

Thread size	Drill size	Thread size	Drill size
⅛ × 40	2.55	5/16 × 40	7.10
5/32 × 40	3.25	5/16 × 32	7.00
3/16 × 40	4.00	⅜ × 40	8.70
7/32 × 40	4.80	⅜ × 32	8.70
¼ × 40	5.50	7/32 × 40	10.30
¼ × 32	5.50	½ × 40	11.90

An earlier chapter described the cutting of an unusual shaped hole, these three photographs show how to do it. First a row of holes is drilled. These are then either sawn to a slot with a coping saw, or piercing saw, or a small chisel is used to join them. The hole is then filed smooth.

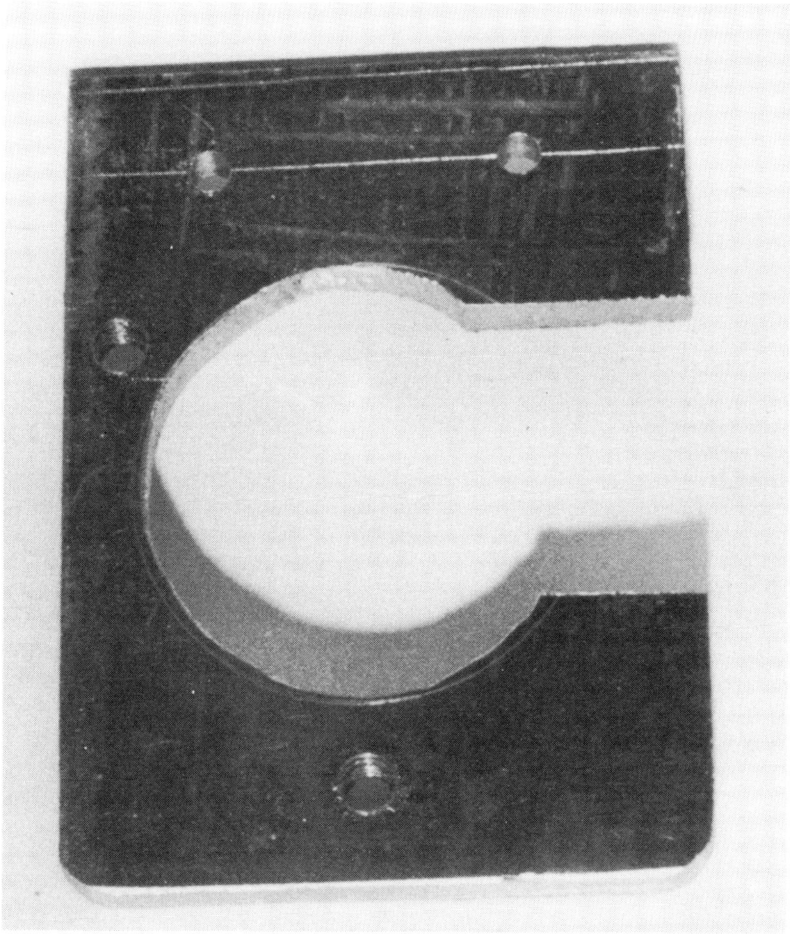

10 *THREADING*

Putting threads on metal falls basically into two categories. Male threading, for which we use a die, and female threads for which a tap is used. The actual operation of tap or die is somewhat similar and there is nothing very difficult about either. In both cases the watchword is care and as long as no attempt is made to hurry the operation there is no reason why there should be any problems. The main difficulty with both types of threading is to keep the thread square, it being remarkably easy to make the thread at an angle, even though this looks at first glance like an impossible thing to do. The other problem is breakages. Not so much with dies, although this can happen, as with taps, particularly in the small sizes.

DIES

It is most improbable that the model engineer will use anything other than a circular die. This is a circular piece of steel in one of a series of standard sizes. Through the centre is a hole with a thread of the size for which the die is intended. There are three or four holes outside this which just break into the threaded one. It is the gaps caused by these holes that help the die to cut. They also create a refuge for the swarf which is created by the die. Usually one of these outer holes will have a small slit in it which goes through to the outside edge. There will probably be two other indentations along the outside edge as well. Dies are made in either carbon steel or high speed steel. The carbon variety is useful, but has a very limited life—half a dozen or so threads cut with it on mild steel and it is of no further use. If kept for brass and similar materials on the other hand they will last quite a long time. If the die is just for an odd job then buy a comparatively cheap carbon one but if it is to be used regularly high speed steel is the one to go for. There is a considerable difference in price, but the high speed one will be worth the money as it will last a long time.

Dies are held in stocks, more commonly just referred to as die holders. These are round pieces of steel, bored to take a particular die size. Two arms are fixed to it to enable it to be turned, and there should really be three screws in it. These fit the slot and the two indentations in the dies. When tightening up the die, the centre one should be tightened first as this opens the die out. The other two can be tightened enough to secure the die in the holder. If the thread is too tight it should have the die run along it again with the centre screw loosened, and the two outer ones tightened up harder. The piece of metal to be threaded should have a slight taper on it to allow the die to pick up. The die must be wound down perfectly square. Once one and a half turns have been made check for squareness, and keep checking after every two revolutions. A tapping compound, which can be purchased at any good tool dealer should be used. The die should be wound off frequently, and the swarf cleaned from it with a brush. An old toothbrush being ideal. If threading to a shoulder, you may find that the die just stops a little short. It can then be turned round and used in the other direction. The correct side to start with is the side with the details printed on it. It is now possible to purchase dies already in holders and these are set in such a way that the thread is formed without adjustment to the die. It is also possible to buy die nuts. These are dies with a hexagon shape. They are mainly intended for cleaning up old threads.

Above—*The normal type of circular die. The side with the manufacturer's name should be used to make the thread as this has a lead to allow it to settle square on the metal. To get close to a shoulder the other side can be used.*

Below—*A die in a stock or holder.*

TAPS

Using taps is a slightly different proposition to using a die. There are three types of taps; taper, second, and plug or bottoming. For perfect threads all three should be used although it is possible to get near perfect threads using just two, and personally I always dispense with the second

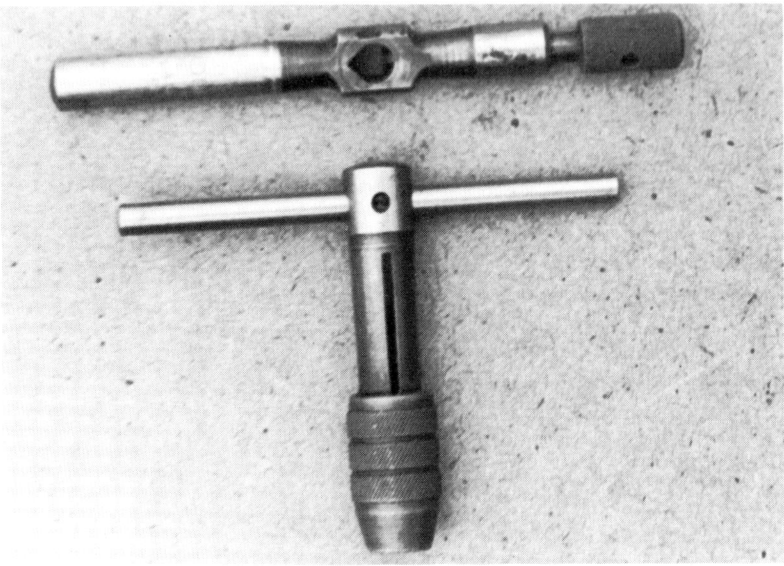

Above—*Two types of tap wrench, at the top the bar and underneath the chuck type.*

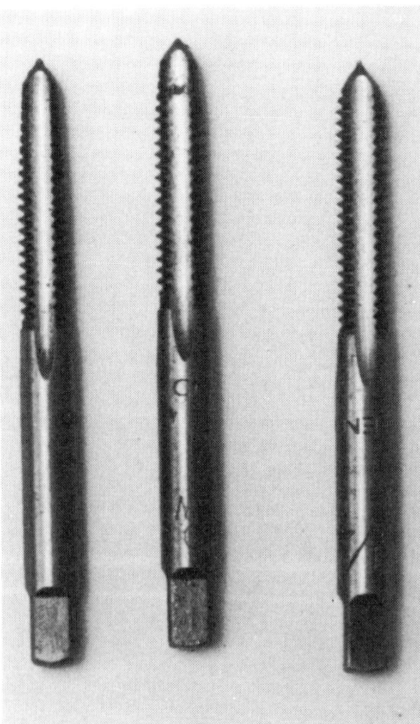

Left—*Three types of tap are available. From left to right they are taper, second and plug.*

Above—*For small taps the chuck type of tap wrench is most useful, it should be used one handed but very tiny taps can also be supported by the shank.*

Below—*Although larger type chuck wrenches are available, the bar type is probably better for heavy work. Two hands should be used on the wrench.*

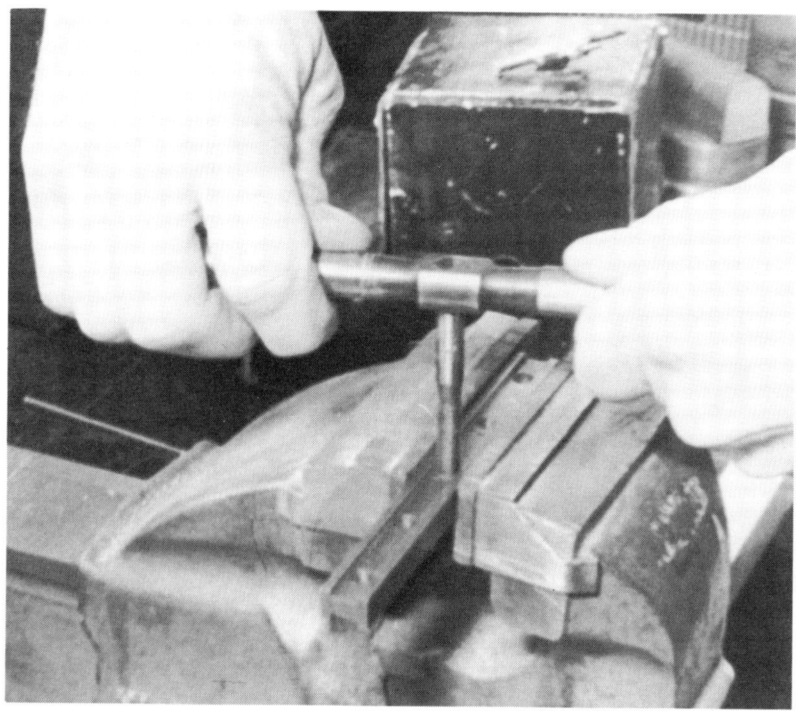

tap. Taps are held in tap wrenches. These come in two shapes; one like two round bars screwed together, the other like a small chuck with a bar across the top. Which one is used is generally a matter of personal preference but very large taps will almost certainly need the bar type. Most of the taps you will use will be either three or four fluted. There are other types but these are meant for special purposes and have little application in model engineering.

TAPPING

Before tapping, the correct size hole must be drilled. A table should be referred to for this. The first tap to use is the taper. This is ground in such a way that it should enter the hole you have drilled before starting to cut a thread. Once the tap is felt to be home turn it gently, making sure it is

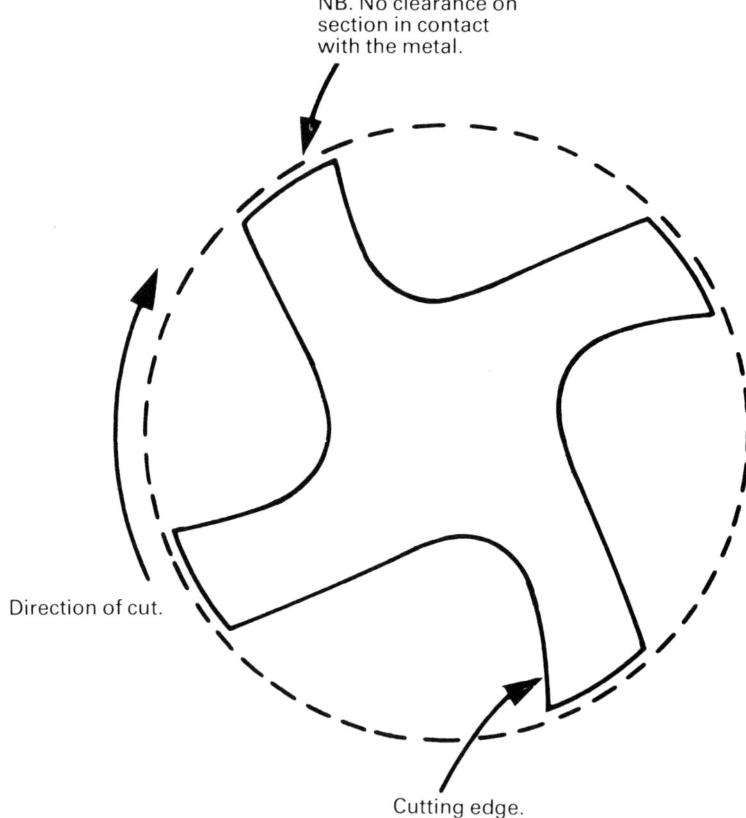

NB. No clearance on
section in contact
with the metal.
Direction of cut.
Cutting edge.

Above—*A sketch showing the main points of tap construction.*

Opposite—*Throughout threading operations check that the tap or die is square. Here a piece of metal is being used, there being insufficient room to use a toolmaker's square.*

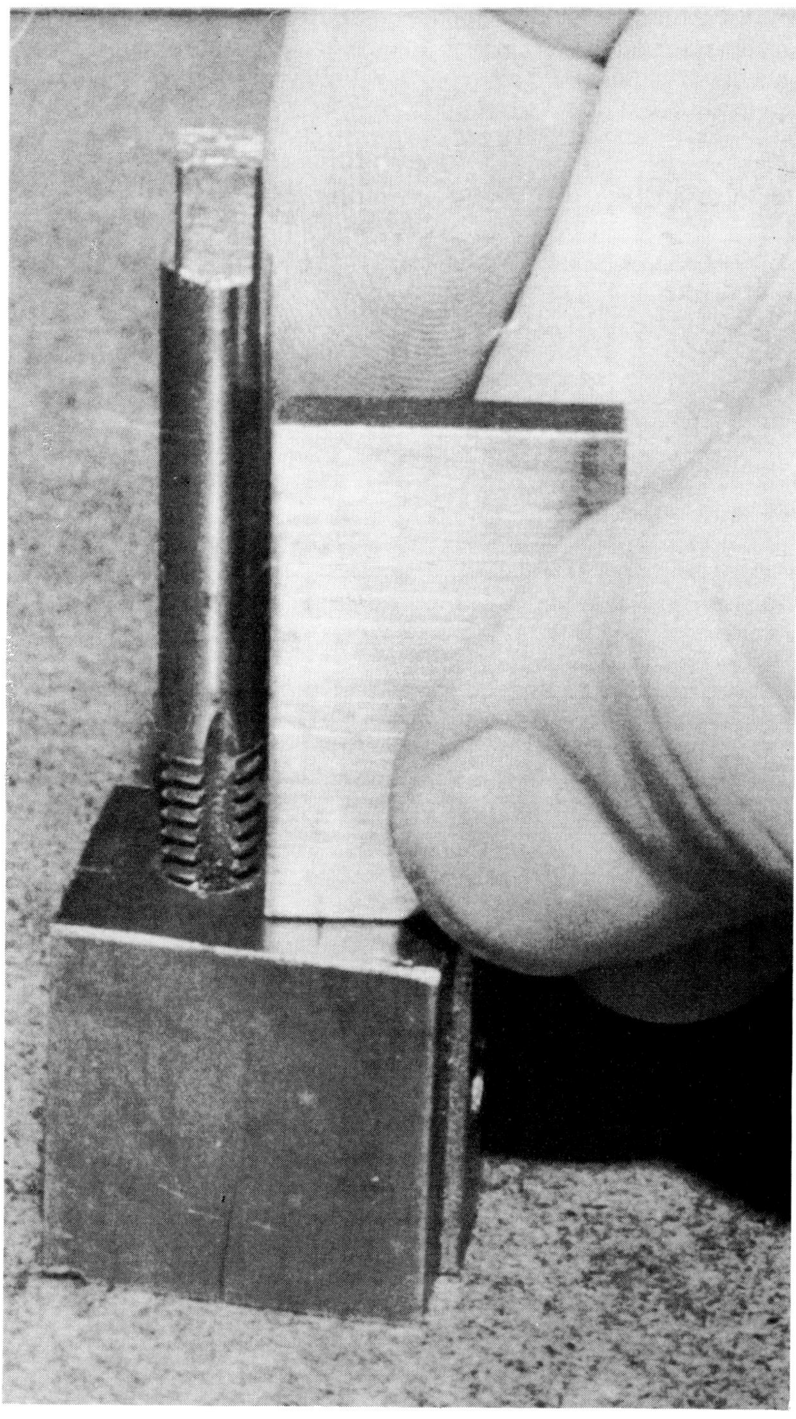

kept at 90°. A tapping compound should be used. Exactly the same rules apply to tapping as to using a die. Remember that if the tap tightens up it must be withdrawn. If it cannot be withdrawn easily it will have to be wiggled a little bit, backwards and forwards, until it frees itself. Be gentle at all costs. Taps are expensive, but more important still it can be very difficult to extract taps from holes after they have broken. Better to be careful from the first.

CAUSES OF BREAKAGES

It is not often that dies will break in use, and if they do it is almost certain to be a combination of the die not being square and being treated too roughly. More likely it is that the metal will break in the die, leaving the job ruined and the die full of unwanted metal that cannot easily be extracted. If this happens then the best thing to do is to drill centrally through the piece of metal left in the die with a small drill, leaving a small tube of metal in the die. Saw through the wall of the metal in two places with a ground down hacksaw blade. It should now fall out.

BROKEN TAPS

Taps break for the same reason as dies and will also break with unsteady handling. Small taps in particular will break with a sudden movement. When they do break extracting them is quite a problem. If the tap has passed through the metal and a piece is left protruding it may be possible to turn this and unwind the tap. However this will only be possible if the tap has been broken by a sudden jerk on the part of the operator. If it has caught up there will be no moving it. If it is a taper tap, then a sharp blow on the taper, if protruding, with a hammer may do the trick. If not it is probably just a case of punching it out and drilling and tapping a larger hole. Taps in blind holes in small sizes are virtually impossible to extract without special equipment. The best answer is to take special care not to break the tap.

THREADS

There are numerous different threads that have been designed for all sorts of purposes. As far as this book is concerned I have concentrated on model engineer series, these being nice fine threads suitable for such purposes. Although in theory British Association (BA) threads should have been extinct a long while ago they are still very much alive, and still quite popular, and have been included in the following tables for this reason. I have also included a few British Standard Pipe Threads. These are needed if fittings are purchased for water piping and they are bought from an ironmonger. All of the many other series that were once available now appear to be on the decline and so they have not been included. There is no table for metric threads. Simply take the pitch away from the diameter and you have the tapping size. For example: Thread 6 mm diameter by 0.5 pitch, shown as 6 × 0.5, take away 0.5 gives a tapping size of 5.5 mm.

BRITISH ASSOCIATION (BA) THREADS TAPPING SIZES

Size	Tapping size	Clearance size	Size	Tapping size	Clearance size
0	5.10 mm	6.10 mm	9	1.55 mm	1.95 mm
1	4.50 mm	5.40 mm	10	1.40 mm	1.75 mm
2	4.00 mm	4.80 mm	11	1.20 mm	1.60 mm
3	3.40 mm	4.20 mm	12	1.05 mm	1.40 mm
4	3.00 mm	3.70 mm	13	0.98 mm	1.30 mm
5	2.65 mm	3.30 mm	14	0.80 mm	1.10 mm
6	2.30 mm	2.90 mm	15	0.70 mm	0.98 mm
7	2.05 mm	2.60 mm	16	0.60 mm	0.88 mm
8	1.80 mm	2.25 mm			

11 *BENDING*

Sooner or later in your modelling the need to bend metal is likely to arise. Without special equipment this can be a somewhat daunting task. It is not too bad with very thin material, but as the metal gets heavier, then the problem increases. One common factor with all bending is that the metal to be bent must be as soft as possible. This apart, there are a few dodges that we can use which will help us in these sort of operations.

SHEET METAL

Probably the type of metal that will require bending most frequently is sheet. Relatively thin sheet can be put between two pieces of angle iron in the vice and bent over, using either another piece of angle or a heavy

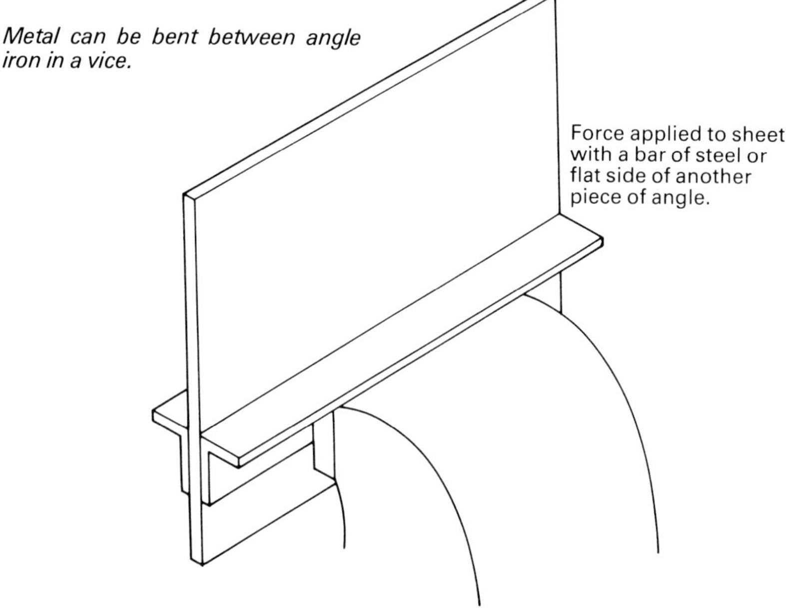

Force applied to sheet with a bar of steel or flat side of another piece of angle.

Metal can be bent between angle iron in a vice.

A bending bar. This consists of a bar of metal which is folded back on itself in such a way that it will spring open when not compressed. The metal to be bent is inserted in the gap between the bar and clamped in a vice. This gives a straight supporting edge, much the same as two pieces of angle iron, which can be used to bend metal accurately using either hand force or a hammer to bend the metal to the required angle. The bending bar, being in one piece, is somewhat easier to use than two pieces of angle which can be difficult to line up.

bar section. This enables the whole length to be bent in one go and prevents distortion. If the metal is too thick to be bent by hand then hammering will have to be resorted to and this will inevitably cause a certain amount of distortion. Should such a method be required, then you must tap the metal over evenly a little at a time. The metal should be supported between pieces of angle and hammer blows should be directed as close to the supported part as possible. If it is possible to heat the metal and bend it whilst hot this will help. Blacksmiths use things called bending bars which serve the same purpose as the angle iron, but being all in one piece are somewhat easier to manipulate. A method that will help in bending thicker sheet metal is to use a piece of an old car tyre, or some very thick rubber, in one side of a vice and a piece of angle in the other. An assistant will be required to line things up, but it is quite effective and with a really hefty vice fairly thick metal can be bent. If the metal to be bent is too long for the vice then the idea can be extended by using a piece of metal plate of the required length behind the rubber. This method does not give a 90° bend, but as the hardest part of bending any metal is the first part, it gives a start that should enable the job to be finished.

Using rubber and angle iron to start a bend.

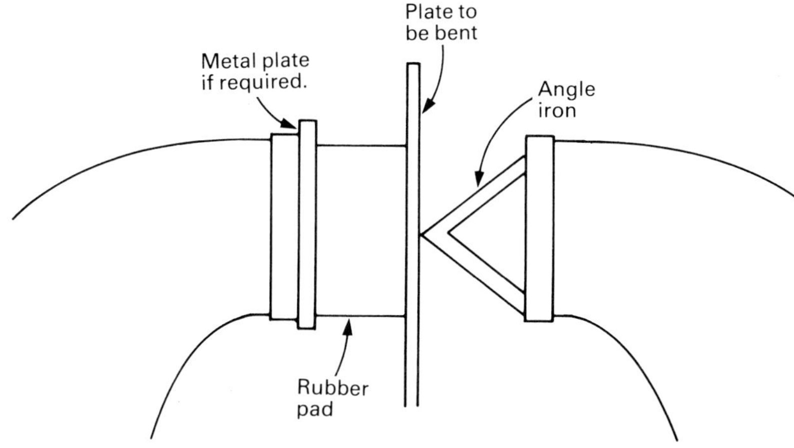

Below—*A sheet metal folder for use in the vice.*

Left—*A floor standing sheet metal folder.*

SHEET METAL FOLDERS

It is possible to purchase sheet metal folders, either of a type that will fit in a vice, or a floor standing model that is considerably larger. Not many home workshops will have room for the larger version, but the small bench type can be quite an asset if a great deal of such work is envisaged. It is also possible to make such a machine, if a little thought is given to the construction.

TUBE

Tube can be bent by a variety of methods. Springs can be obtained that fit either inside or outside the tube to support it during bending, preventing kinking. Bending tube without any form of support will always end up with a kink on the inside of the bend. A simple method is to use a device with two wheels which have been recessed to take the tube, usually with varying sizes of wheels to accept different size tubes, which will do the job quite nicely. Again it is easy to make such a device and the effort will repay itself over and over again. One other method that used to be quite popular but which seems to have lost favour of late, is to soft solder a plate on the end of the tube, fill it with dry sand (it must be dry), and solder a plate on the other end. You now have what is in effect a tube with a movable core. The sand is flexible enough to allow the tube to bend, but solid enough to prevent any kinks.

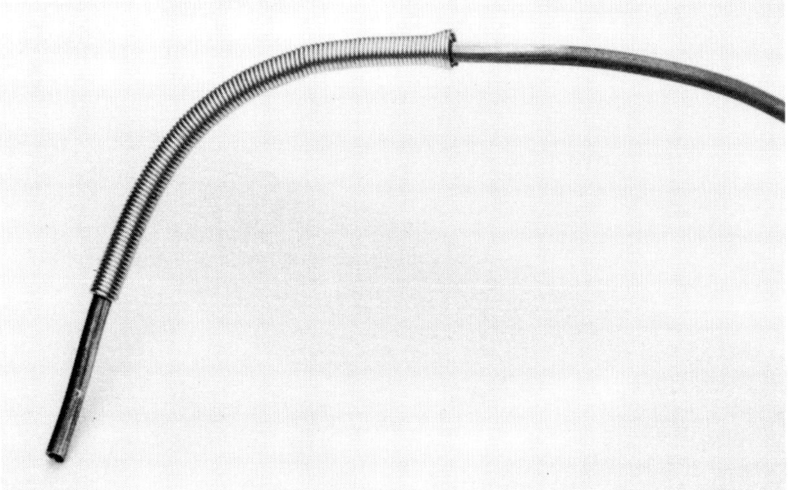

Above—*Tube can be bent in a spring. Internal springs are also available.*

Below—*Using a home made tube bender.*

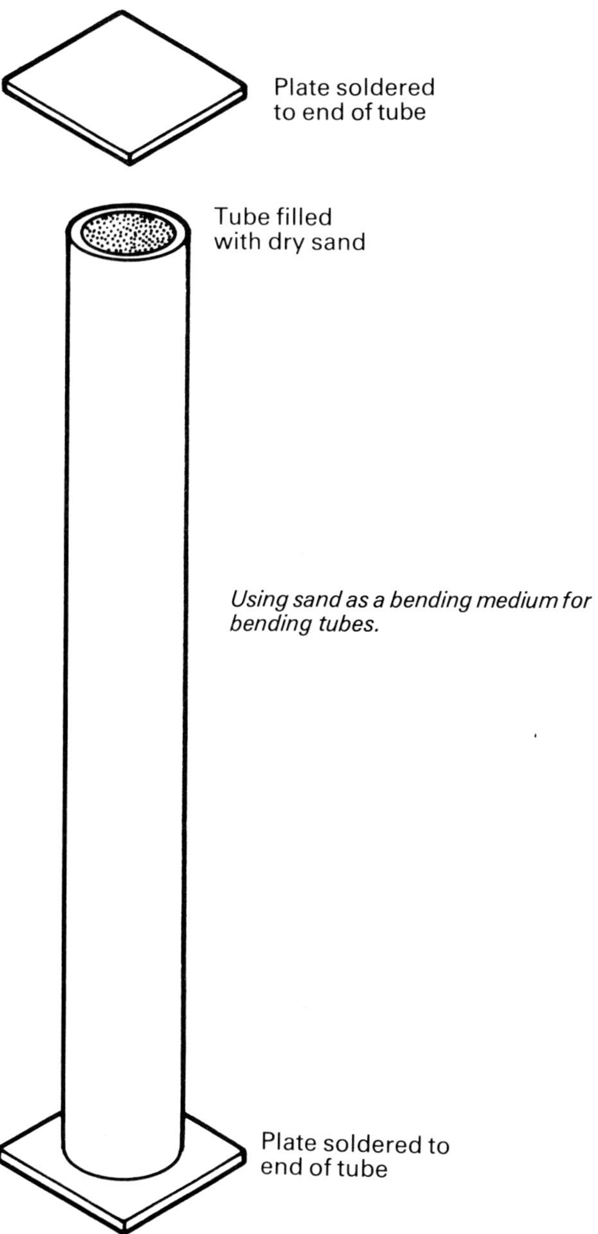

Using sand as a bending medium for bending tubes.

BAR

The only way to bend bar is to put it in a vice and use brute force. If at the time of bending it is red hot then less of that brute force will be required. Remember if possible to have the piece you are using as a lever as long as possible.

ROLLING

If a tube or similar shape of object has to be made from sheet, then the metal will have to be rolled. Rollers are made for this purpose, usually consisting of three metal rollers, possibly geared together. The sheet is supported on two, and the third applies pressure, making the required radius. If rollers are not available, thin sheet can be rolled round metal bars or wooden formers. If possible the metal should be softened first. It is then best to start by laying the sheet of metal on something like a piece of foam rubber of reasonable thickness. Roll the former backwards and forwards over the metal which will start to curl. It should then be drawn round the former using stout wire or jubilee clips. It is not too difficult—I have rolled a boiler shell for a 5 in gauge locomotive round an old piece of drainpipe and saddle tanks for another one round several pieces of hardwood which I glued together and then turned round on a lathe.

Bending rolls are ideal for forming round work. These are larger than average, rolls are available for use in the vice.

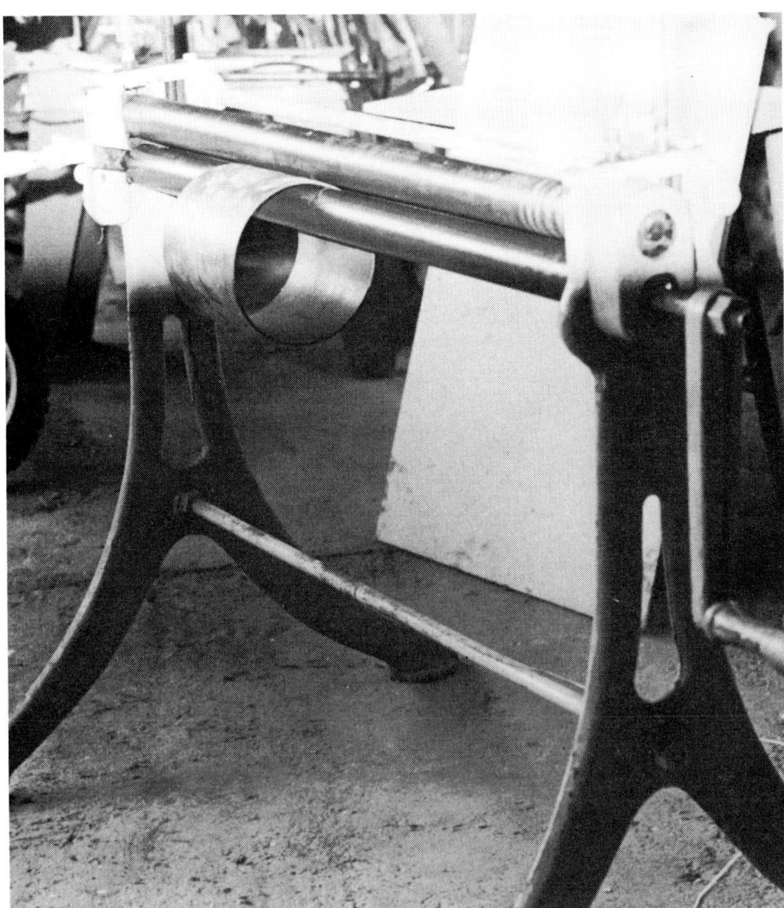

Above— *The need arose to roll a large length of brass, and bending rolls were not available. This wooden former was turned from sections of wood glued together and the brass rolled round it.*

Below— *The piece of brass which was rolled round the wood. The curve is quite perfect, with no distortion.*

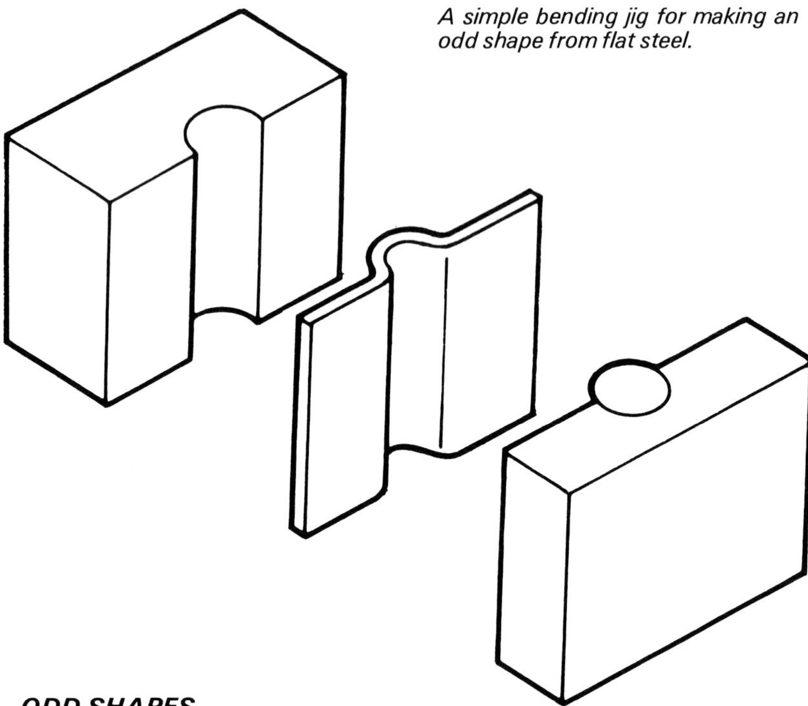

A simple bending jig for making an odd shape from flat steel.

ODD SHAPES

Sometimes odd shaped sections and brackets are required. These can be made by making up simple little bending tools and using them in a vice. The tools need not be elaborate, and as they will probably only be used once or twice they can be quite rough. A male and female section will be needed, and do not forget to make allowances for the metal that will have to go in them.

BENDING ALLOWANCES

When metal is bent through 90°, the outside surface becomes stretched and the inside compresses. It is therefore necessary to make allowance for this when bending is done, otherwise the bend will not be in the correct place. If we think a little about these facts it becomes obvious that somewhere, about half way through the thickness of the metal is a line (unseen of course) where the metal neither stretches or compresses. Because there is a slight difference between the compressive strain and the tensile, or stretched, strain the line is not as we would imagine, along the centre, but somewhere nearer the inside. Now there are all sorts of complicated formulae for working this out but as it depends on the type and thickness of material it would be impossible to give all the required figures and so I must generalise. If you take the line as being as 0.4 times the thickness of the metal from the inside radius then you will not be too far out. As the bend referred to is a right angle, or as near as you can get it, if you add 0.4 of the thickness of the metal being bent to the length of the part you should end up about right.

In the case of bends that are not right angles you can still use the right angle formula of 0.4 thickness, but in this case it is multiplied by the radius of the inside of the bend. As the radius gets larger, and in the case of tube, then the figure can be calculated as 0.5 of the outside diameter of the tube or the thickness of the metal. Let me point out though that such calculations are only approximate, probably sufficient for most model engineereing purposes, but if you are trying to get to within say a hundredth of a millimetre, then a lot more information will be needed.

Producing tapered tubes is a matter of working out the required length of metal at each end. The large end will then have to be marked centrally at a point where a right angle from each edge would meet. If a curve is now drawn through this mark to the end of the metal, the end of the cone will be square when rolled.

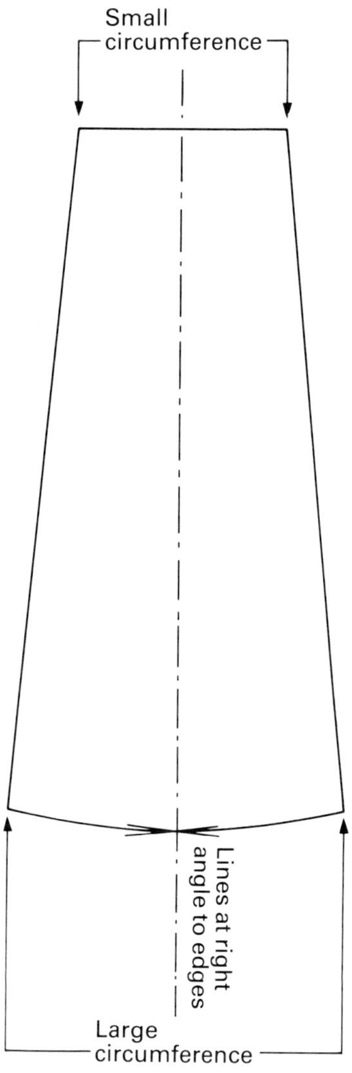

To make a tapered tube draw lines at right angles to edges on centre line. Where they cross will give centre of curve required.

12 *CLAMPS*

Clamps are used for several reasons when working with metal, for example, holding work together for drilling through two or more components at once, holding it together for riveting, brazing and welding, and also for some marking out. In addition some small work that is being sawn or filed may at times be held more conveniently in a clamp than in the vice. It follows then that various clamps will form an essential part of your equipment and it is worth building up a stock of different types and sizes as and when the opportunity occurs. Clamps that are used to hold work together for brazing and soldering rapidly deteriorate with the rusting action of the flux and, whilst it is possible to keep them reasonably clean, if a lot of such work is to be carried out separate clamps should be kept for this purpose.

TOOLMAKER'S CLAMPS

Toolmaker's clamps will probably be the most useful to you. These consist of two bars of square metal with two screws passing through them in opposite directions. The result of this is that by adjusting the two screws until the bars are parallel it is possible to not only get the clamp to line the work up accurately, but it will also cover a larger area of metal than do most other types of clamps, making it grip the metal really tightly.

Incidentally, before describing other types of clamps, there is an anomaly here in that clamps used for woodwork purposes are largely called cramps. There is no real difference, and as far as this book is concerned the word clamp will be used throughout. If purchasing them, however, it may be as well to bear in mind the fact that the shop assistant may know the item you are asking for as a cramp, rather than a clamp.

Right— *The standard type of toolmaker's clamp is available in a range of sizes.*

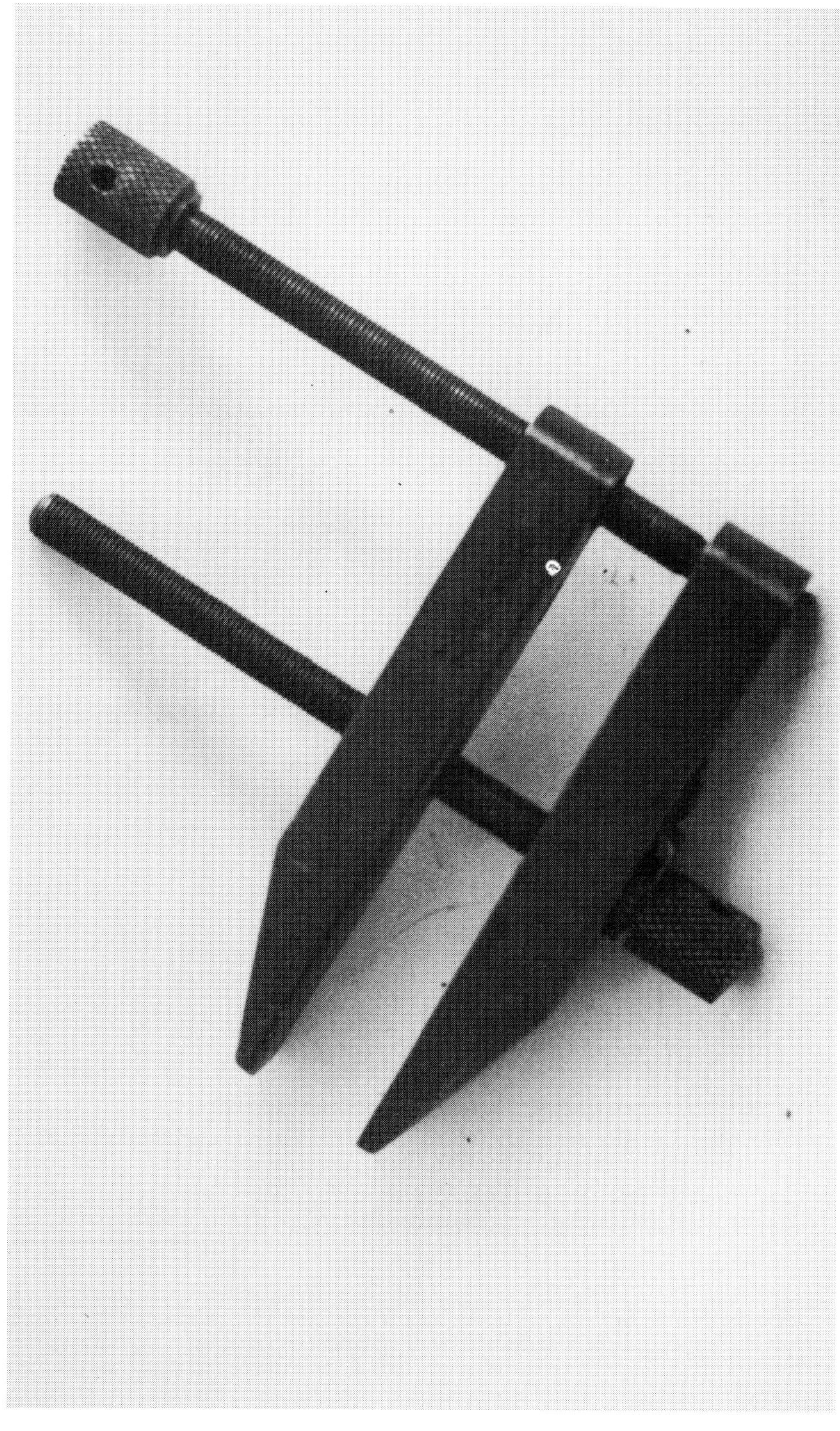

'G' CLAMPS

You will also find 'G' clamps, or patented versions of them, useful. A 'G' clamp consists of a frame in the form of a flat horseshoe. Through one end is a screw and on the end of that screw a loose piece of metal to actually fit on the work and so prevent the screw either biting into it or slipping off as it is tightened up. These types of clamps are extremely useful for larger work and for holding work to the bench or a machine. However they usually need to be used in pairs, as a single 'G' clamp does not have enough purchase to hold work steady. A single one will often suffice, however, for holding work to be brazed or welded, and the use of such a single clamp means that work can be released more easily than with a toolmakers clamp.

A simple 'G' clamp.

LONG CLAMPS

Sometimes you will need clamps of considerable length, and then neither the toolmaker's clamp, nor the 'G' clamp, will do. A woodworker would probably use something known as a sash cramp in these circumstances but these are not suitable for metal working and one of the patent types of clamp on the market will be needed. Many of these work on a principal of sliding the part with the screw fixing along a bar, and then a lever action takes over when the clamp is tightened up. They are very useful for the metal worker. Another type consists of two sections of clamp that slide on to a metal strip. Again tightening is at one end only. The advantage of this type of clamp is that different lengths of bars can be used for different work, giving a great deal of flexibility. One other advantage to this particular type is that the actual clamping ends hold interchangeable pieces, designed for different types of work. Included amongst these are items like small vee blocks which are ideal for holding round work.

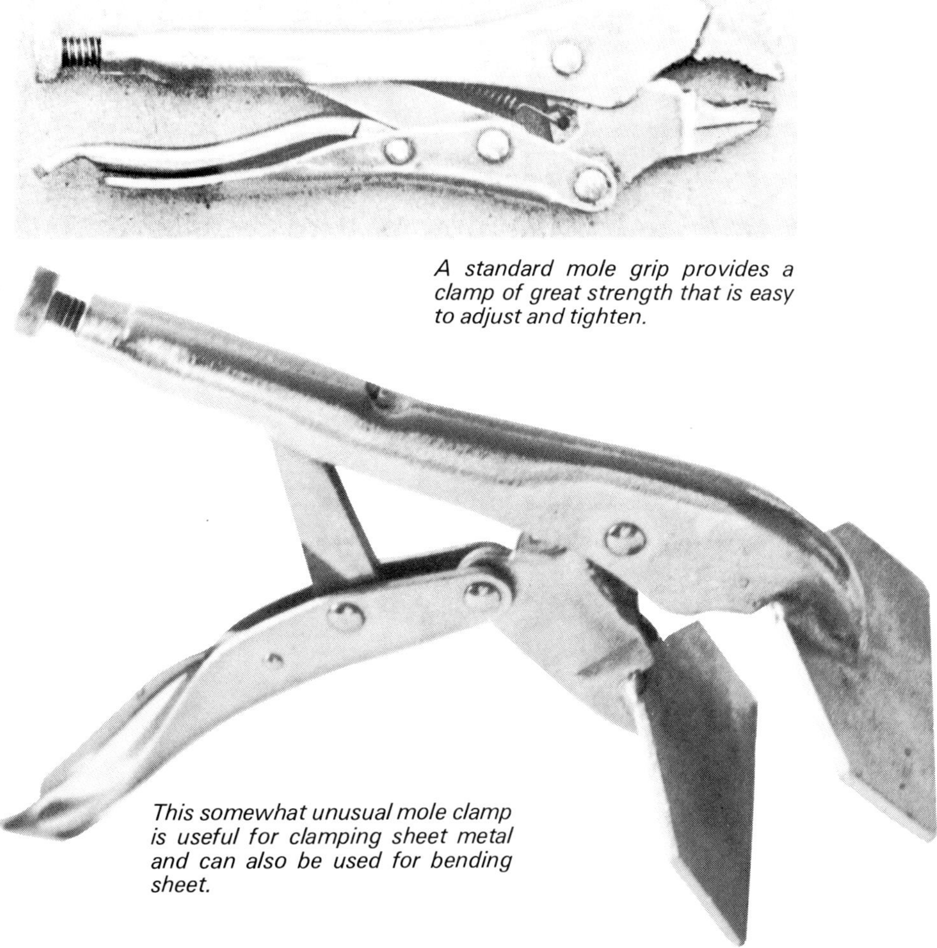

A standard mole grip provides a clamp of great strength that is easy to adjust and tighten.

This somewhat unusual mole clamp is useful for clamping sheet metal and can also be used for bending sheet.

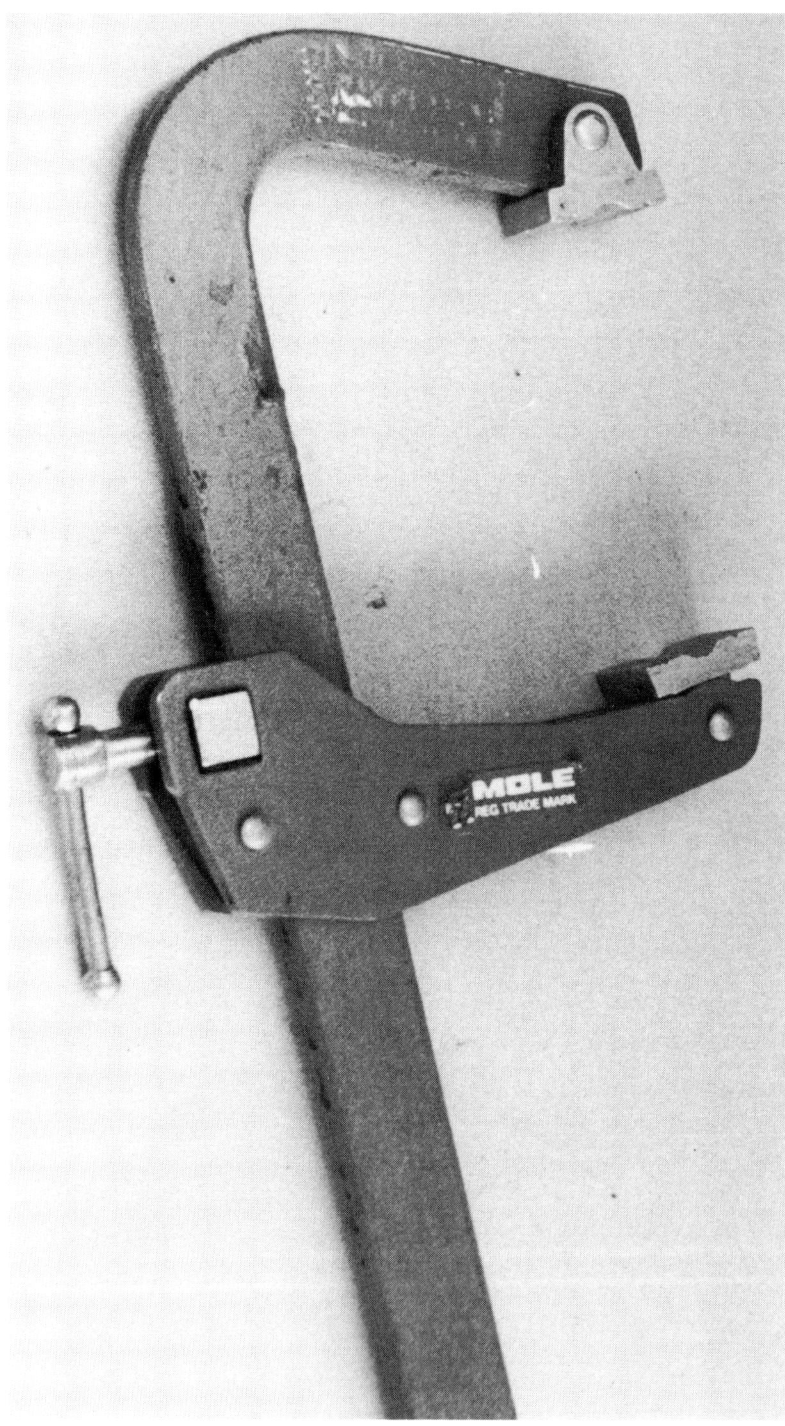
MOLE
REG TRADE MARK

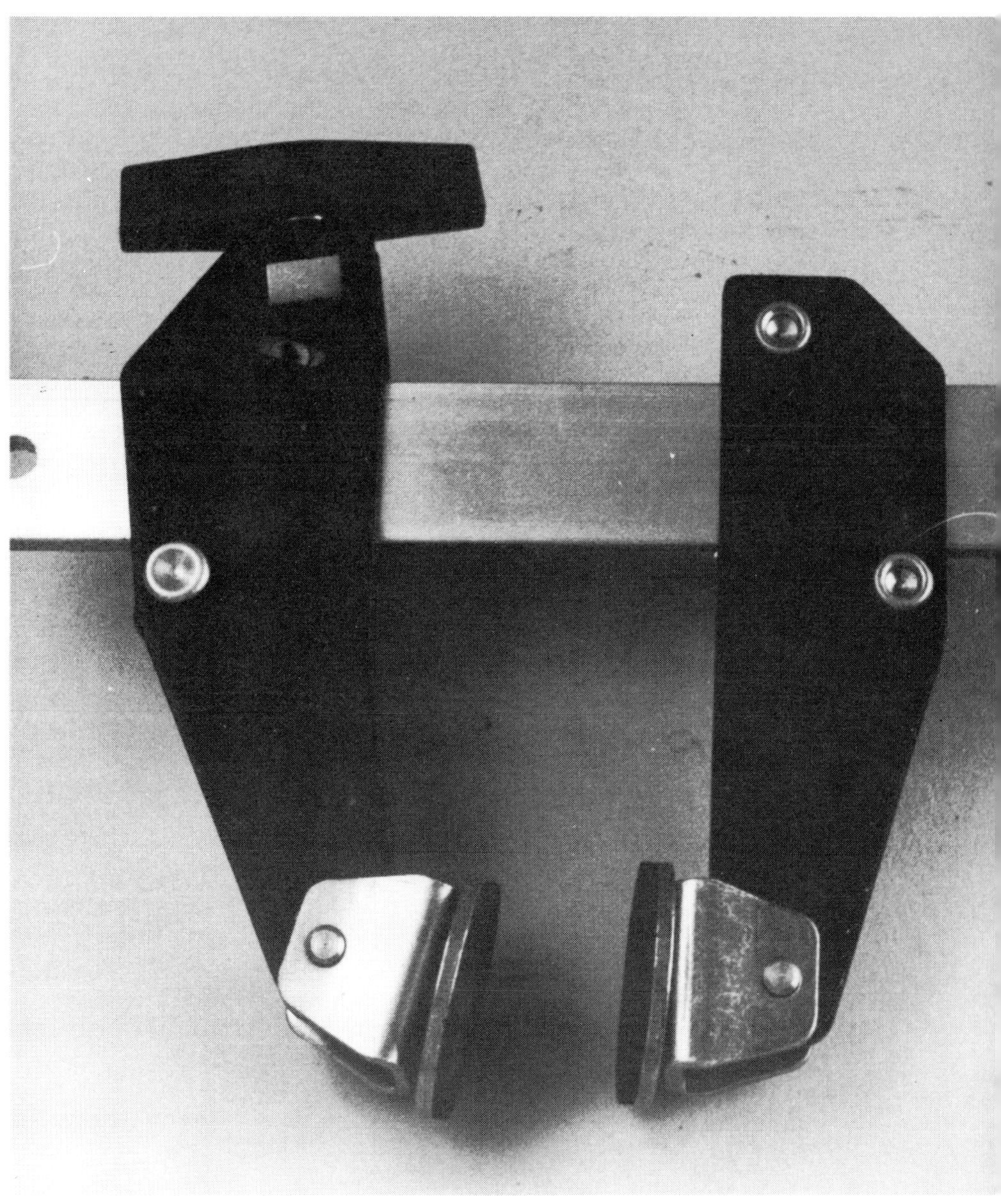

Above—*A jet clamp, useful because it is easy to tighten and a range of ends are available for different types of work.*

Opposite—*A mole rack clamp, useful for long work.*

CLAMPS FOR SMALL WORK

Where very small work is to be filed or drilled it can be secured in a toolmaker's clamp and the clamp held in the vice, or even in the hand, to work on it. This means that there is no question of the work being damaged by excessive pressure. There are some devices on the market designed for this purpose, including one referred to as a hand vice, which has been on the market for many years now, and would always form part of the tool kit of any good bench fitter. It is really only a clamp but it can either be held in the hand or in a vice, once the work has been set in it. These hand vices come in a variety of sizes, but one with jaws about 50 mm across should suffice for most of the work that a model engineer will undertake.

Left—*A hand vice, useful for holding small parts together as well as for a small vice to hold work with for filing etc.*

Below—*Using a toolmaker's clamp to hold together a small locomotive frame.*

IMPROVISING

The question of improvised clamps should not be ignored. Frequently a standard clamp proves to be unsuitable in certain circumstances. A few bars of metal with holes drilled in them, and some studding (metal rod threaded throughout its length) or some long bolts, together with a supply of nuts will more often than not take care of clamping up work in awkward situations.

When clamping work, whether with orthodox clamps or with the improvised ones, do not forget to protect the metal before tightening up the clamps if the finish is to be important. If used properly toolmaker's clamps do not cause any great problems but other types can do considerable damage. Pieces of thin wood, or even card, placed in the right position will prevent a great deal of effort in trying to get clamp marks out. I have seen work completely ruined by careless clamping. Included in my own tool box for the purpose of protecting my work, are some small pieces of felt, and small pieces of an old cycle tyre. The walls of the tyre are ideal. These have now been augmented by some soft plastic cut from an old ice cream tub of the type sold for the home freezer. This material does not slip and has the advantage of being very flexible.

13 *THE HAMMER*

It may sound a little silly to devote a chapter to the hammer—everybody knows a hammer when they see one. I once heard it described as a stick of wood with a steel end on it. It lives on the floor just under the bench or at the bottom of the tool box. I suppose that really sums up just how much thought we give to this tool, which in fact we will probably use more than any other. In point of fact hammers are tools with their own personality, and it is as well to know their individual characteristics.

It is probable that your tool kit will hold more than one hammer. Over the years I have picked up about eight, and all are used quite frequently. Let us then think about the types that will prove most useful. You will certainly use one for centre punching and also for riveting. One will be required for bending metal, and possibly one for adjusting work by giving it a smart but light tap. Fortunately for all these purposes we can get away with two types, a ball pein hammer and a soft faced hammer.

The ball pein hammer has the normal hammer head on one side with a shape resembling a half ball on the other. It is ideal for centre punching and for riveting, for which you need a light one weighing no more than ¾ lb. This will enable you to tap lightly on the punch and to start the riveting process slowly. For finishing off the riveting you will need a larger hammer of about 1 ½ lb or even 2 lb. This size will also be useful when you need to bend any metal.

Soft faced hammers come in a variety of types. The idea of the soft face is as one would expect to prevent marking the work. Although hammers can be bought with plastic and nylon heads, the best for our purpose is one with a copper head on one side and a hide one on the other. These heads will usually unscrew so that they can be replaced as they wear. They are sold not by weight but by diameter of the face of the head. A 1 ¼ in or 1 ½ in should be suitable for most jobs.

Make sure that the face of a hammer is kept clean and shiny. If it gets marked, the marks on the head will transfer to the work when used on sheet metal. Hammers can easily be polished with an abrasive cloth. You should also make sure that the head is held tightly to the shaft.

Loose hammer heads are not only dangerous but they may move during working and spoil the work by causing the blow to be struck in the wrong place. Always hold the hammer at the end of the shaft, not near the head. Holding it this way enables you to give a better swing and to get a better feel to the blow. Feeling the blow is important in order to judge the amount of power.

Soft faced hammers are used in the same way. The heads should be kept trimmed and not allowed to become splayed out. The hide can be trimmed with a knife, and the copper with a file. Make sure the heads are well screwed in when in use.

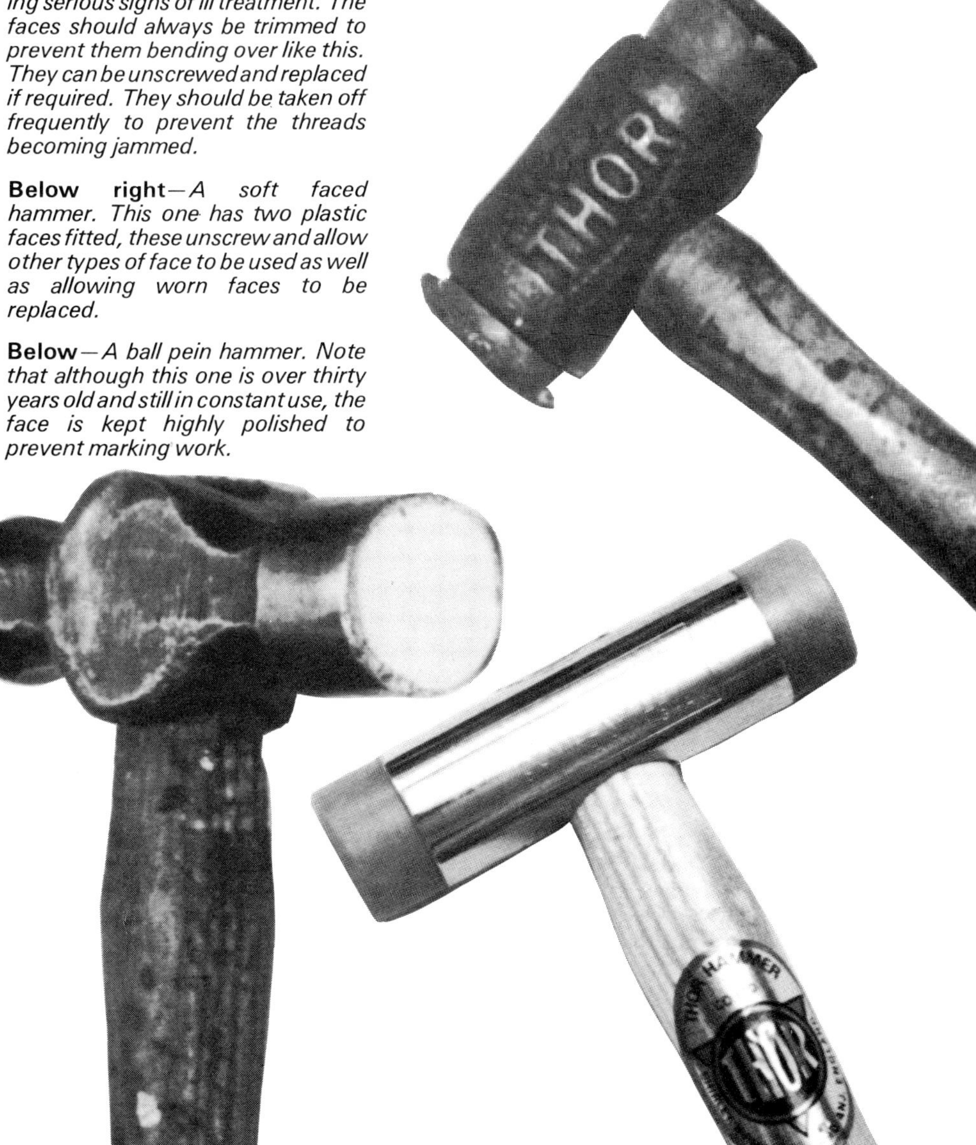

Right—*A hammer with copper at one end and hide at the other, showing serious signs of ill treatment. The faces should always be trimmed to prevent them bending over like this. They can be unscrewed and replaced if required. They should be taken off frequently to prevent the threads becoming jammed.*

Below right—*A soft faced hammer. This one has two plastic faces fitted, these unscrew and allow other types of face to be used as well as allowing worn faces to be replaced.*

Below—*A ball pein hammer. Note that although this one is over thirty years old and still in constant use, the face is kept highly polished to prevent marking work.*

14 *SOLDERING, BRAZING AND WELDING*

There is invariably, when engaged in modelling, a necessity to join metal parts together. Both parts may be of the same metal or they may be two different kinds. They can be screwed, bolted, riveted or glued together but there will also be times when you will need to use either soldering, brazing or welding. Soldering and brazing really boil down to the same thing, with only the temperature used in the process differing. What happens is that another metal of a lower melting point is stuck to the metals to be joined by melting it on to them. If they are perfectly clean then the metal which is melted, which is called solder or braze, sticks to them. Thus if two objects are in close contact when the solder or braze is applied to them a join will be formed. It sounds very simple but as always there are certain rules that need to be observed when making the joint. The most important of these is that the metal object receiving the solder or braze must be heated to a temperature where the joining metal will melt on it and flow along it. There are more bad soldered and brazed joints caused through attempting to melt the solder and applying that to the metal than for any other reason. The solder or braze must melt on the object to which it is being applied, this simple fact cannot be stressed enough and failure to observe it will mean a poor joint. The other important factor to be remembered is that the place where the joint is to be made must be clean and a suitable flux must be applied that will prevent oxides forming on the metals being joined.

Before proceeding with a description of soldering and brazing it will be as well to explain the difference between these and welding. Here I am only concerned with the type of welding that can normally be carried out in the home workshop, in industry other processes are available which give more flexibility. The main difference between welding and soldering is that whilst in soldering the solder is used to make the joint, when welding the two objects are raised to such a temperature as to cause the metal of which they are made to melt and flow together so that we end up really with one object, rather than two pieces joined together. Metal is used in welding as a filler, but it takes no part in the actual joining process at all. This means that we are somewhat limited in the metals that we can weld at home, and in fact the choice more or less comes down to bright or black mild steel and cast iron, and even cast iron needs special treatment. Nevertheless welding is an operation worth consideration.

OXY-ACETYLINE

There are two methods of welding, the first involving the use of oxygen and acetyline gases, always known as oxy-acetyline. Bottles of the two gases are required and these have to be fitted with pressure gauges and hosepipes which lead to a torch that will mix the two gases. The acetyline is lit first and the flame adjusted until there is just a slight sign of black smoke leaving the end. Oxygen is then turned on to approximately an equal volume. When adjusted the flame should have a blue cone, this is the hottest part and is what gives the desired result. The sketches should give a reasonable idea of what is required, but if you are to do any amount of welding it might be as well to attend your local evening institute, who in my experience run some first class welding courses. The amount of heat applied will depend on the size of the nozzle used and the amount the gas taps are released by. A certain degree of experience will be required to get things right. Too small a nozzle for the thickness of metal will result in a poor weld, too large and the metal will dissolve before your very eyes.

Once the flame has been correctly adjusted welding can begin. The metal being welded must be brought to a point where molten pools can

Three types of gas flames with an Oxy-acetyline torch.

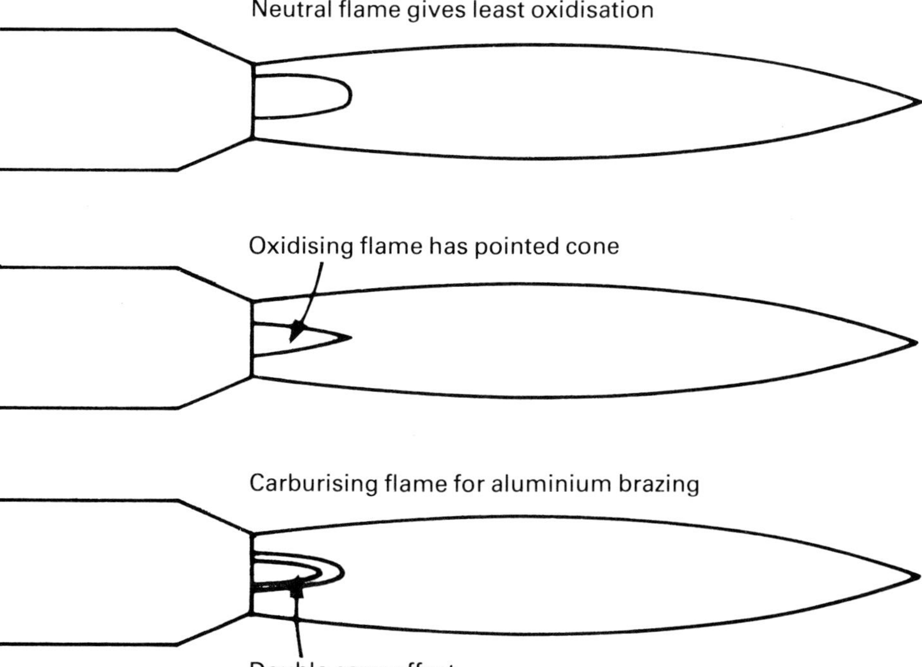

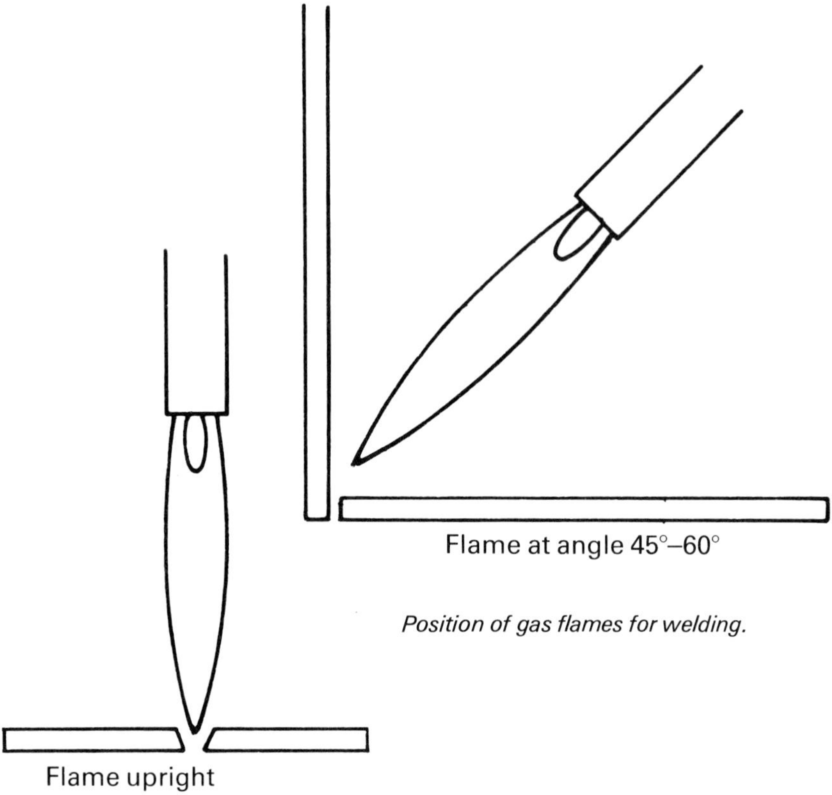

Flame at angle 45°–60°

Position of gas flames for welding.

Flame upright

be seen. Filler rod is then applied to fill up the molten metal. Do not under any circumstances try to apply filler rod as you would solder, it is not intended for this purpose. Metal that is to be welded together should have a slight recess filed along the edge and a small gap may be left. The torch should be held at a 60° angle when welding butt joints, but held upright for edge welds. The sketches will explain far better than words possibly could.

Dark glasses must be worn for welding.

Whether gas or electric welding the metal should be chamfered and the weld built up in between.

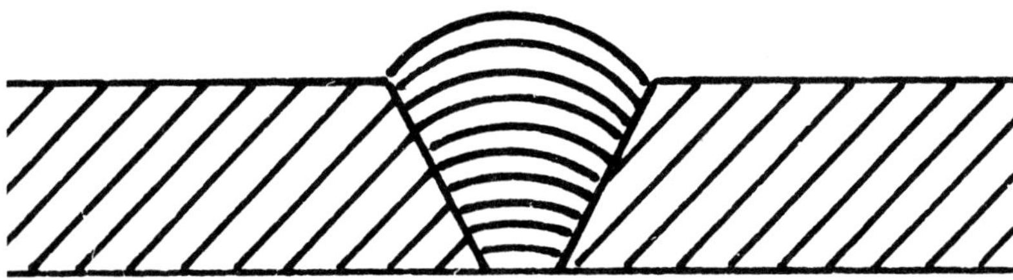

ARC WELDING

Arc welding means connecting a high current of electricity which is of negative polarity to the piece to be welded, and connecting a specially prepared rod to a positive current. As the two are brought nearly together an arc of intense heat shoots across, and causes the metal rod and the surrounding metal to melt, thus giving a weld. It is a quick and easy method and arc welding transformers for home use are now available quite cheaply. The current used will depend on the thickness of the metal to be welded. All transformers are variable, but when buying one you must make sure that you get one of sufficient power. A transformer of 100 amps will only weld metal about 3 mm thick. If you plan to weld thicker metal you will need a heavier transformer. For home use about 140 amps seems to be a reasonable size.

Right—*A typical welding transformer for home use.*

Below—*The earthing clamp and electrode holder of the welding set.*

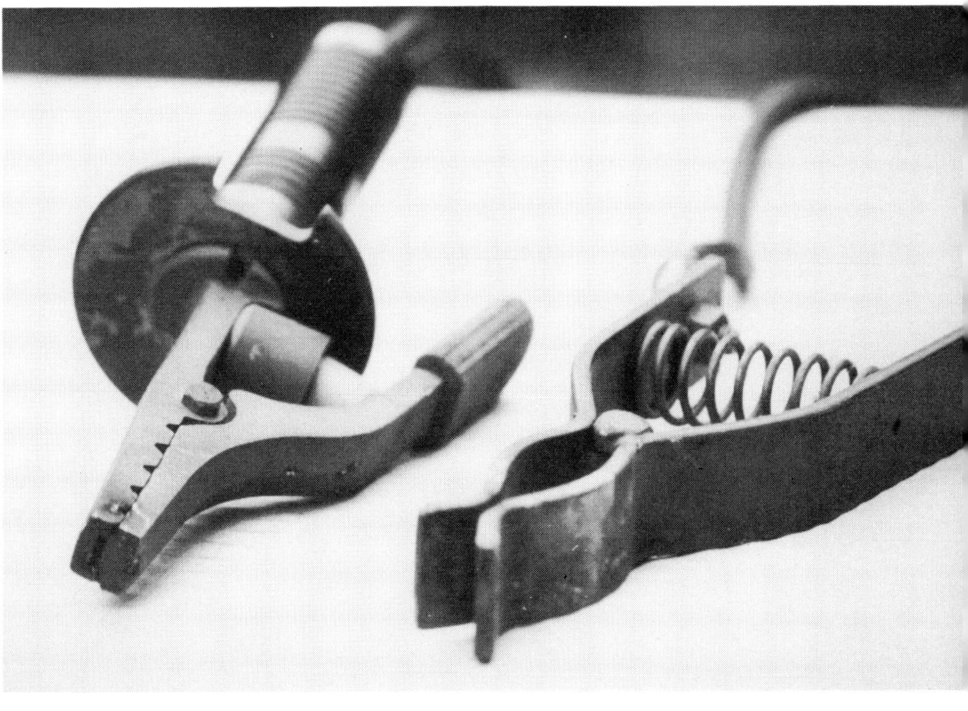

When welding the rod must be brought into the work very slowly, otherwise it will stick. A mask will have to be worn which will protect your face from pieces that may fly off. It will also enable you to get very close to the work, something which is a necessity because of the need to see through very dark glass. Do not under any circumstances arc weld without a suitable mask—it will permanently damage your eyesight. Special welding rods are available to weld cast iron, and sometimes welding rods are marked as negative, in which case the work must be attached to the positive terminal.

A face shield must be used for arc welding.

SOFT SOLDERING

Having discussed welding at some length, let us turn to soldering and brazing. The dividing line between the two is somewhat hazy, as soldering with silver solder is often called brazing, but I will break them up into three different sections, starting with soft solder. All solders are alloys and those referred to as soft solders will contain mostly tin and lead. As these metals both melt at low temperatures so will the soft solder, the melting point being between 190°C and 225°C. These are the type sold in hardware shops and properly used will give a good joint to brass, copper or steel. It is possible to purchase solders that will melt at much lower temperatures, and model makers frequently use these on white metal kits where the uses of a normal soft solder cannot be considered because the melting temperature is too near that of the kit itself. Soft solder is usually purchased in the form of wire or sticks, some of which have a core of flux in them.

FLUXES

If the solder is not ready fluxed, and for many purposes a non-fluxed solder is probably more suitable, then a flux must be applied so you will need to know a little about the types of fluxes and their uses. This also applies to silver soldering and brazing. The flux is basically to prevent the metal, which has already been cleaned, from oxidising. If the metal is perfectly clean and it does not change in any way when heated, then in theory no flux is required. In fact though, the metal discolours and a scale forms on it, that scale may be so thin that we cannot see it, but it is there and it is called oxidisation, the same basic principal as rusting. Flux applies a coat over the metal and prevents that happening. Flux must therefore be capable of two things; firstly of preventing the oxides from forming and secondly of doing so at the temperature at which the solder will be used. Fluxes can be purchased under various brand names for the purpose, either as a paste or as a liquid. The most common of these being spirits of salts, which is sold under various brand names. These are usually effective up to about 300°C. Something that I use, and which I personally prefer to any other, is phosphoric acid. This is sold as a rust preventer under the brand name of Jenolite, and possibly other brand names as well. It is a most efficient flux and I have found it effective for much higher temperatures. Fluxes should be applied to the work before heating, and possibly during soldering operations. They must be washed off after use to prevent corrosion. With ready cored solder no other flux should be required—the flux from the core melts as soon as heat is applied, thus flowing over the joint. Such fluxes are not suitable for high temperatures.

APPLYING THE SOLDER

I have already pointed out how joints should be thoroughly cleaned before heating, and how they should be fluxed. Heat can be applied either by means of a soldering iron or a small blowlamp. Either way the solder must be applied to the metal to be joined. Frequently one sees people trying to solder by melting the solder on to the iron instead of the metal. This can only end in disaster, giving what is known as a dry joint where a layer of solder has built up on to the metal without actually binding to it. At first it looks like a good joint, but after a short period of use will just fall apart. This brings me to the next point, which is that all joints should be tested for strength when finished.

Soldering irons should be tinned before use—this means cleaning the end until the copper shows bright, heating up and then fluxing. It is then coated with solder, any excess brushed off. This tinning helps the solder to flow. Soldering irons come in different sizes and the iron must be big enough to impart the required heat to the joint. A small iron is of little use on large work. The advantage of a soldering iron over a blowlamp is the fact that joints can be made much neater, any surplus solder being picked up on the iron. It is possible to purchase both electric irons and those to be heated by gas. The latter can often be obtained in larger sizes than electric ones. Special ends can be bought for small blowlamps too, which enable them to be used as an iron. These should be treated with care—blowlamps work on the principal of vapour from the top entering the nozzle, if they are tipped forwards liquid gas can flow through the nozzle and catch fire, so it may be as well to think twice about using a solder iron point on the blowlamp. Electric irons whilst hot and switched on should be laid on to some metal object that will prevent them from overheating, known as a heat sink, the object should be capable of drawing off surplus heat. Failure to do this will result in the heating element of the iron burning out.

Three types of soldering iron, all electrically operated. At the top a general purpose iron rated at 35 watts. Centre, an iron suitable for light electrical work. At the bottom a heavy iron rated at 85 watts for use on thicker material and materials that conduct heat away quickly. The iron was supplied with two ends or bits, a large one which is somewhat bulky but retains heat well and the one shown that applies considerable heat to a small area.

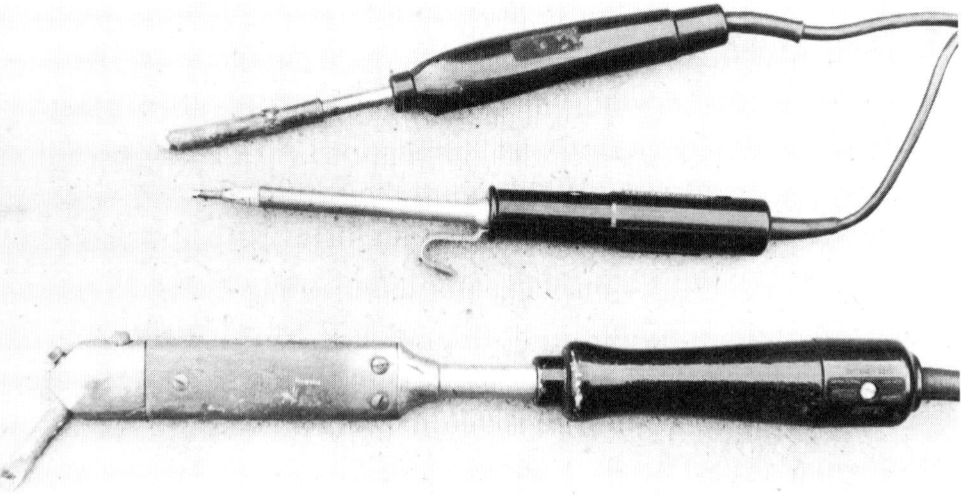

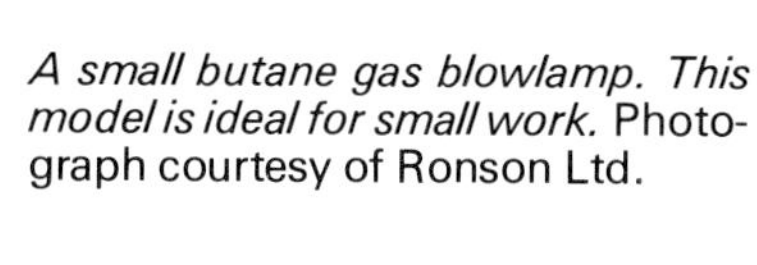

A small butane gas blowlamp. This model is ideal for small work. Photograph courtesy of Ronson Ltd.

A somewhat old fashioned paraffin blowlamp. It will give more heat for its size than the modern liquid gas type but is more trouble in use.

A gas operated blowlamp that will tackle most jobs in the workshop except where a great deal of heat is required, as in boilermaking.

IN BETWEEN SOLDERS

Silver solders usually melt at a temperature of 650°C or more. This leaves a wide gap between these and soft solders. During the last decade or so a number of solders that will melt in between these two have appeared on the market, together with fluxes suitable for use with them. They form a useful addition to the range and as strength appears to go with temperature, they are much stronger than the soft solders. It is probably as well to refer to them by brand name, although different manufacturers will have ranges available under other names. Plumbsol by Johnson Mathey and a similar product by Sheffield Smelting Company called ST 300 are solders which melt in the 221°C–235°C range. They are alloys containing small quantities of silver, as do most of these higher temperature solders. If we go up to between 270°C–285°C then there is a choice between HT5 and SX250. A little higher at 290°C is Comsol, and also in this area melting at between 280°C and 290°C is LM15. Although special fluxes are available, I have found that the solders of these types which I have used will work very well with phosphoric acid. At these temperatures it is doubtful if an iron could be used, no matter how large, and a blowlamp will definitely be needed to get effective results.

SILVER SOLDERS

The next step up is silver or hard soldering in the traditional manner. Here again a wide range of products, melting at different temperatures, are available. The advantage in work of this range is the fact that where several items have to be joined you can start by soldering the first two with a high temperature solder, then for the next piece use a lower temperature, and so on. In this way you are not in danger of unsoldering the parts that have already been joined. The highest of the range that I know of is Silver Flo 16 which has a temperature range of 700°C–830°C. Coming down the scale Silver Flo 24 has a range of 740°C–780°C. At this point I had better say that the reason for the two temperatures quoted is not that the manufacturers do not know the point at which the solder melts. The lowest one is known as the solidus and the higher, the liquidus. Being alloys of various metals, the differing components will melt at different temperatures. When the lowest one melts the solder is in the state of solidus, and when all the components have melted it is the liquidus. So really as far as you are concerned only the higher of the two is relevant. Another solder within this range is known as LX8 and has a temperature range of 700°C–780°C.

In order to use these very high temperature solders a special flux is made. It is called Tenacity No 5 by Johnson Mathey, and Flux F by Sheffield Smelting Company. All silver solder fluxes are based on Borax and are available in powder form which can be mixed into a paste with water or methylated spirits. The advantage of using methylated spirit being that it evaporates leaving the flux residue cleaner after the soldering operation is completed. When made into a paste the flux can be stored in airtight bottles or jars without loss of efficiency. It can also be purchased in tubes of ready made paste.

The better known range of silver solders tend to melt in the 600°C temperature range. Easy Flo and MX20 have a temperature range of 620°C—630°C, and Easy Flo No 2 and MX 12 608°C—620°C. It is quite surprising how much easier it is to work in the lower temperature range than it is even 20° higher.

One problem with silver solders is that usually they will not form fillets, or fill gaps, and special solders are made for this purpose. Again a progressive range of solders for different temperatures is available. At the top is Argo Bond at 616°C—735°C. Next Argo Swift at 607°C—685°C, and finally Argo Flo at 595°C—630°C. There are also a couple of solders available that do not require fluxes—known as Sil Fos and Sibralloy, they will bond copper without flux. The tendency is for them to run everywhere unless strictly controlled, and also to be somewhat brittle after use. They can be useful in circumstances where cleaning metal is difficult as they are self cleaning.

BRAZING

When it comes to brazing you will be working at a slightly higher temperature range to soldering. The flux used can be a commercial brand, or ordinary borax will do. The solder is known as spelter and is really a brass alloy. The temperature required to melt it seems to vary with the manufacturer. Whatever the make, being an alloy similar to brass, the melting point is too near the melting point of brass to attempt to use it on this metal and joints on brass should be silver soldered. It can however be used successfully on steel or copper.

With large propane and butane cylinders a regulator will be needed. The photograph shows the simplest type.

OXYGEN O

ALUMINIUM

Solder and flux for joining aluminium can be purchased, but it is a difficult operation. Aluminium will oxidise very quickly, making it difficult to get solder to stick. Of necessity the solder must be an alloy of a similar type to aluminium and this means there is not a very great difference in melting point between the two, creating the danger, when using a blowlamp, of melting the work as well as the solder. In industry argon gas is used in front of the brazing flame which prevents oxidisation and enables temperature to be controlled much more easily. If you do decide to solder then you will find that the joint tends to be a little more lumpy than an ordinary soldered joint, but good strength can be obtained with care.

BUTANE AND PROPANE BURNERS

For all high temperature soldering some means of retaining heat will be required. Most home workshops are equipped with propane or butane gas in a simple blowlamp. Whilst the high temperature required can be

Opposite— *Using a propane/oxygen torch for brazing small fittings.*

Below— *These are the burner heads, the larger the head the greater the quantity of heat available.* Photograph courtesy Primus Sievert Ltd.

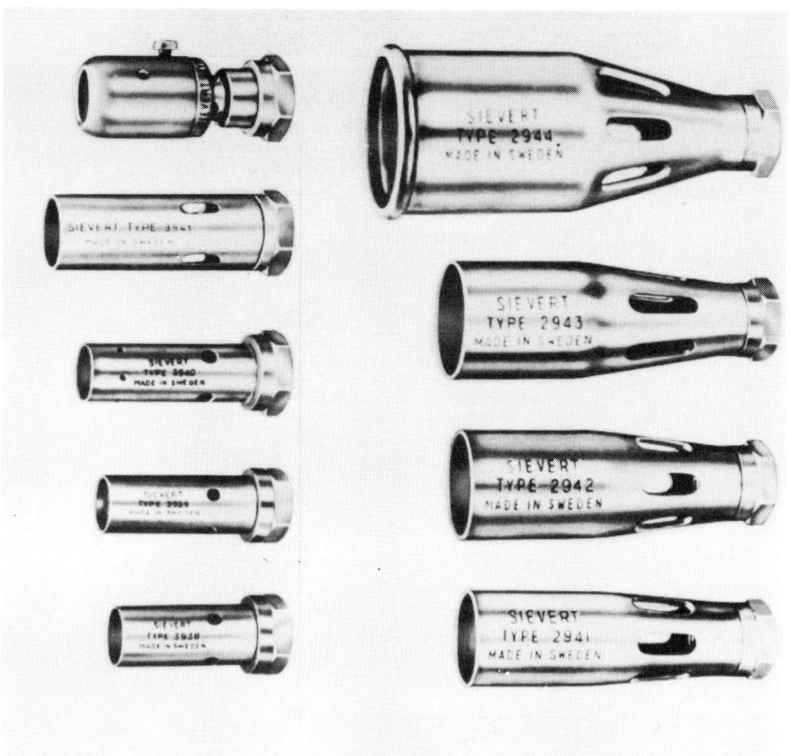

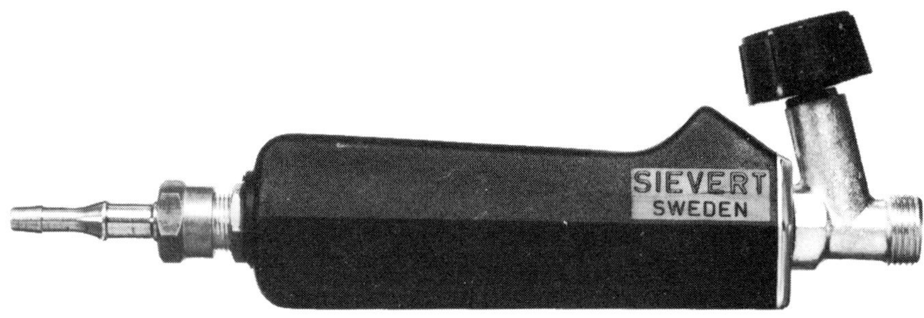

A typical handle for torches operated from large cylinders. Photograph courtesy Primus Sievert Ltd.

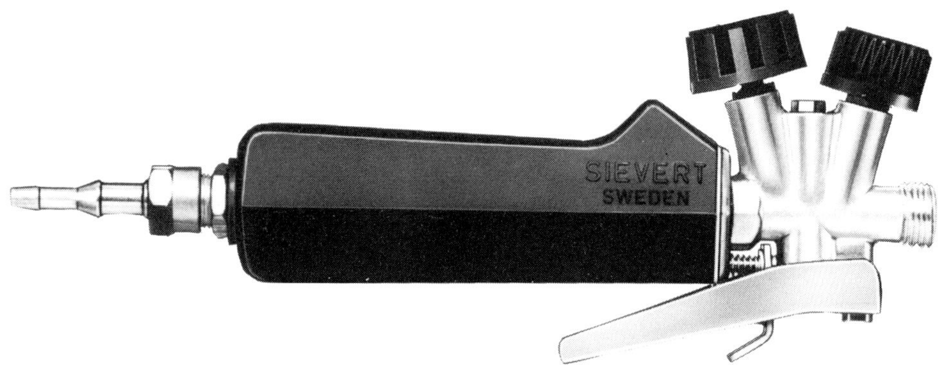

A slightly more advanced handle for a blowlamp which has separate gas and air controls. Photograph courtesy Primus Sievert Ltd.

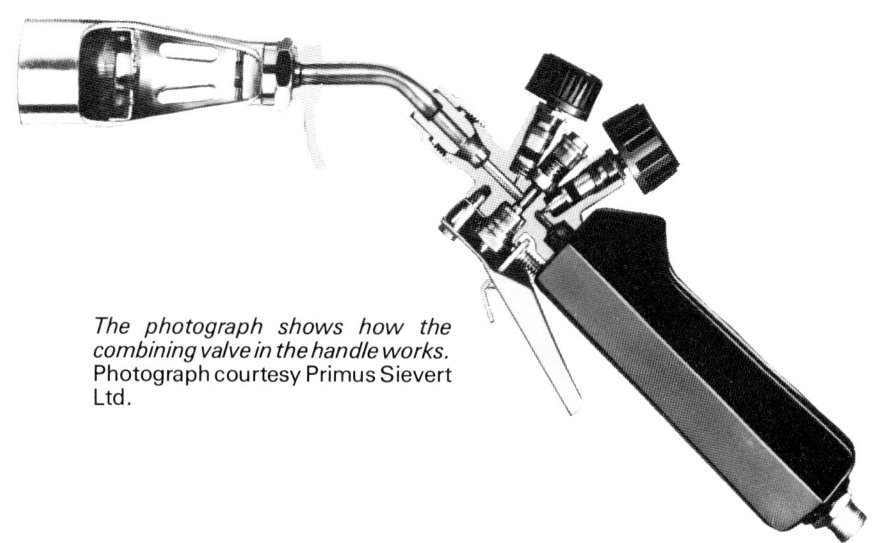

The photograph shows how the combining valve in the handle works. Photograph courtesy Primus Sievert Ltd.

reached there is considerable heat loss, and this must be minimised. Packing the work with coke can help, or making a small brick cavern for the work to go into. I have a small bench specially made up of angle iron and lined with firebrick which I find very useful. A variety of torches and burners are available for use with propane and butane gas. Oxy-acetyline concentrates the heat more, and equipment can now be purchased that is suitable for the home workshop. At one time only the large industrial types of bottles were made and these could not be obtained by the home worker. A quantity of acid is useful for cleaning metal. This should be either sulphuric or hydrochloric, sulphuric being the better of the two. Acetic acid can be used but this needs considerably longer to work, and it may mean leaving pieces in it for several days at a time. If left uncovered in the workshop acid will give off fumes that are corrosive to tools so this is not advisable. Any form of soldering, soft or hard, will create fumes both from the flux and the solder. These are both corrosive to tools and a potential health hazard. It may be worth while wearing a mask which can be obtained very cheaply and can save many health problems. Also make sure when using bottled gas that it is turned off at the bottle after use.

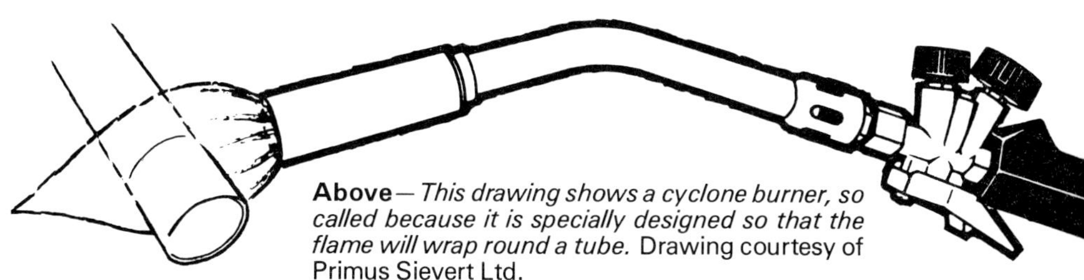

Above— *This drawing shows a cyclone burner, so called because it is specially designed so that the flame will wrap round a tube.* Drawing courtesy of Primus Sievert Ltd.

A propane burner in action. The backing firebrick reflects heat back on the work, saving gas and time required to heat material. Photograph courtesy of Primus Sievert Ltd.

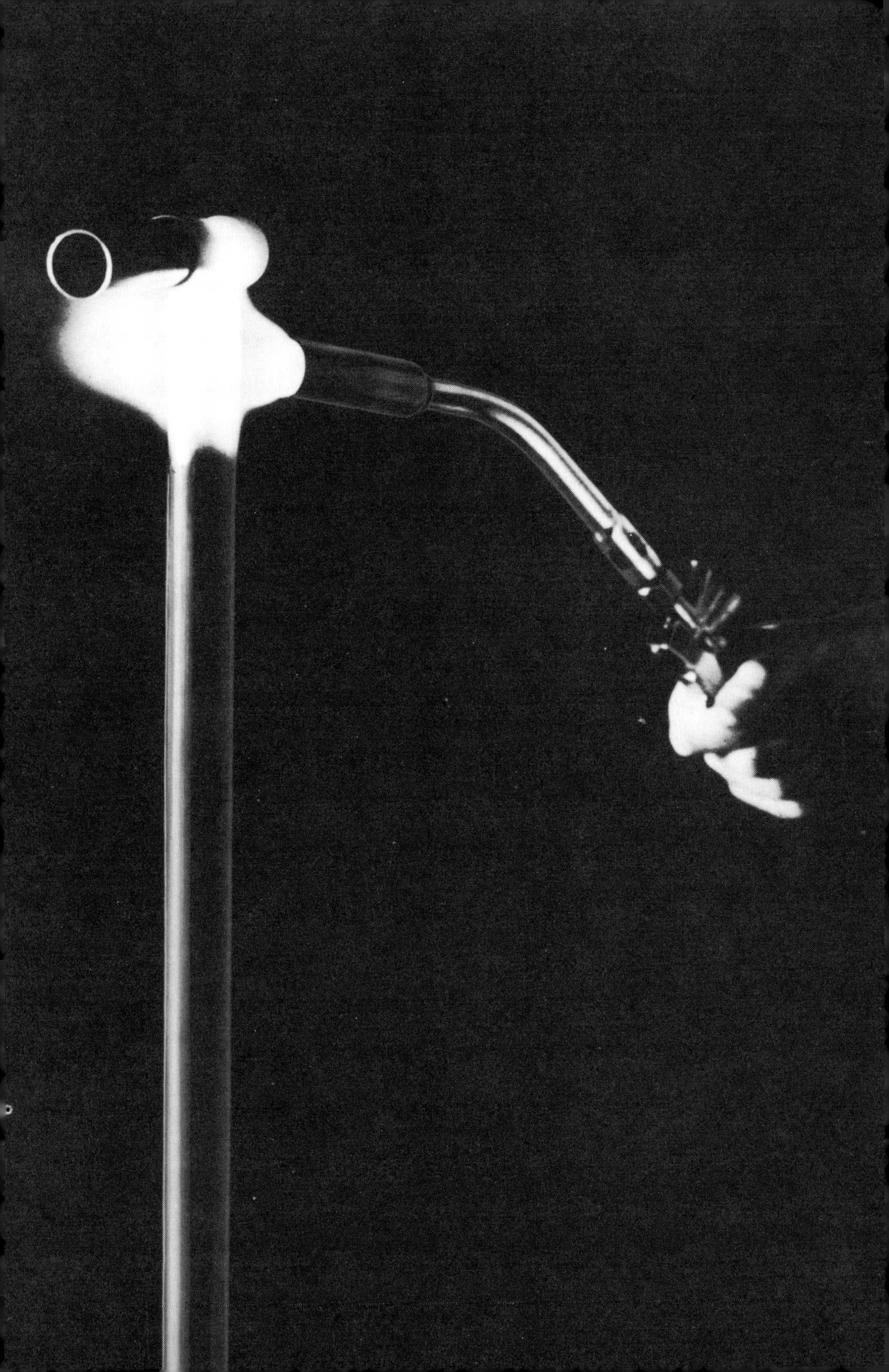

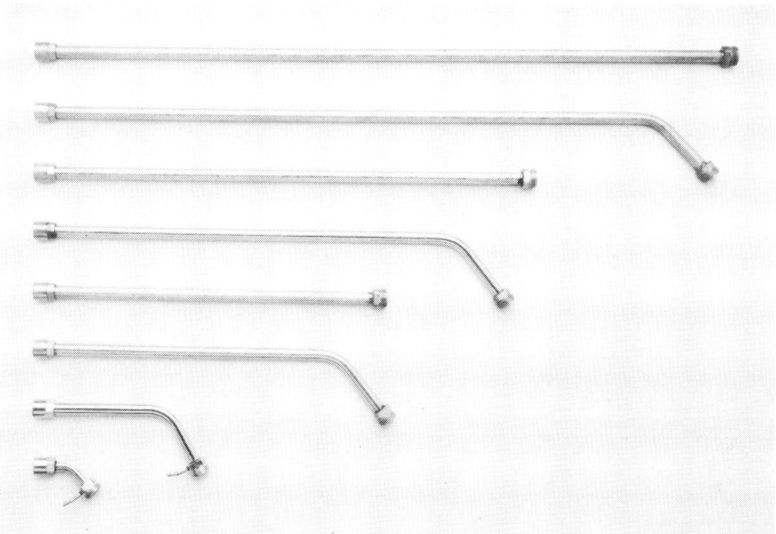

Opposite— *The cyclone burner in action.* Photograph courtesy Primus Sievert Ltd.

Above— *These tubes go on the burner handle and connect to the burner itself. The large range enables one to pick a tube of suitable length for the work being done at the time.*

PURCHASING SILVER SOLDER AND SPELTER

Silver solder can be purchased in a variety of forms. The most usual way being in stick either 1.5 mm or 3 mm diameter. It is also available in flat strip. It can be obtained ready fluxed in rod form. It is possible to buy it as a fine wire, which is particularly suitable for small work, and also as a thin flat foil or washers. The advantage of buying it in this form is that it will be able to use the minimum possible quantity when joining two flat surfaces. It can also be purchased as ready fluxed paste. This being a mixture of ground up solder and a flux. Again a very desirable idea for very small joints. Spelter is available in rods, of 1.5 mm and 3 mm diameter as well as in wire form. Welding rods for both gas and electric welding are available in a variety of thicknesses, the larger the job, the thicker the rod to be used. They are also available designed for different types of material, so check that you get the correct ones for the job to be done.

15 *RIVETING*

Riveting forms quite an important part of the skills required to be an efficient model engineer, no matter what type of modelling you are involved in. Most of the things that you are likely to model were riveted together in full size and in order to make models appear authentic they too will probably have to be riveted. Do not go overboard though, as too many rivets will look worse than too few. Equally the rivets that are used must be to scale, at least as far as the head which shows is concerned. One sees far too many models with oversize rivets and to my eyes they look ghastly. Also not all of the prototypes, particularly as far as locomotives were concerned, had row after row of rivets showing. Many of the designers thought that the showing of rivets meant the showing of bad workmanship. The rivets had to be used but the countersunk side was the part open to viewing, and when this was properly finished and painted there would be no sign of the rivets whatever.

AUTHENTICITY

Rivets are available with various heads but of these, as far as the modeller is concerned usually the round head and the countersunk ones are those required. When making a model try to get photographs which show sufficient detail to allow you to be really authentic. If the original had flat head or pan head rivets then the model should also have them, and this advice also applies to nuts and bolts. Never use anything that would not have been on the original. It is highly doubtful whether many engineering masterpieces would have been assembled using slotted screws with six inch diameter heads and so the models should also be made without the use of large slotted screws. Nuts, and sometimes, bolts, were often square headed. These can be purchased or are easy enough to make, so if the original had square headed nuts and bolts then so should the model.

RIVETS

A rivet is a piece of soft metal with a head on one end and a round shank. The shank is hammered over after the rivet has been passed through

A typical rivet snap for forming rivet heads.

adjacent holes in two pieces of metal, and this holds them together. Rivets are made in a variety of metals. In industry, more often than not, iron ones were used — they were put in whilst red hot, which both made them easier to hammer into shape and also had the advantage that the rivet would shrink when it cooled, thus drawing the riveted parts closer together. There is no reason whatever why the model engineer should not use rivets that are red hot, although usually they are put in cold.

Rivets are put into the holes and then a tool known as a set put over them. A set is merely a piece of steel with a hole in it of the same diameter as the rivet shank. The metal plates being riveted together are then closed up by tapping them with the set, whilst the rivet head is supported. The rivet can then be closed. First of all the protruding part is knocked over to the rough shape required by use of the ball on a ball pein hammer. A tool called a snap is then used to finish it off, and give a good finish to the job. The rivet section to be hammered over should be one and a half times the diameter of the rivet shank. The distance between rivets should also be one and a half times the diameter. When rivets are countersunk, after hammering flat they should be filed or ground flush to the work. A normal countersink invariably leaves a thin gap between the rivet and the metal. This is unsightly and has to be filled. It is possible though to avoid this by using a drill for the countersink. Take it in just a little further than the point so that a slight lip is seen. Now when the rivet is finished off that gap will be missing.

On the left, combined rivet set and snaps, on the right, rivet snaps.

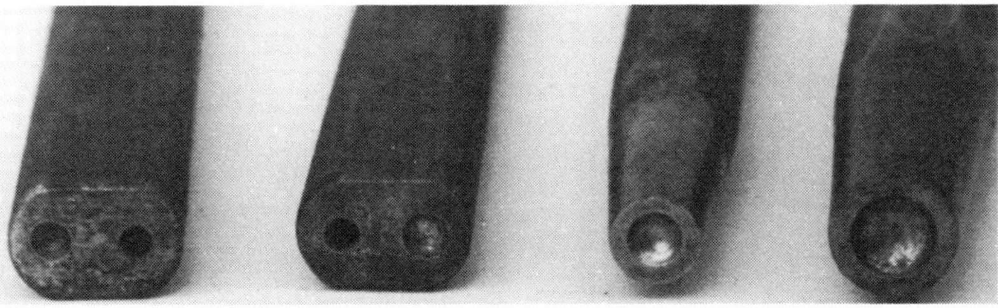

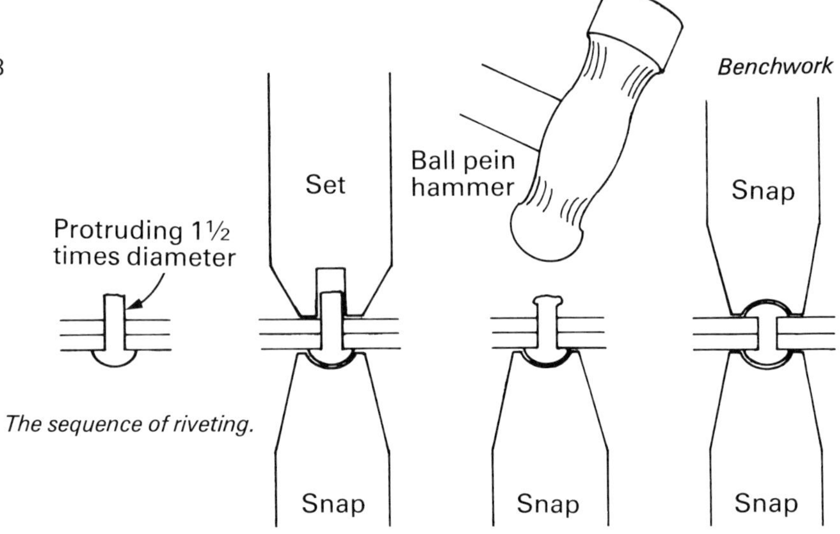

The sequence of riveting.

Below—*The reason for using a drill instead of a countersink when closing countersunk rivets.*

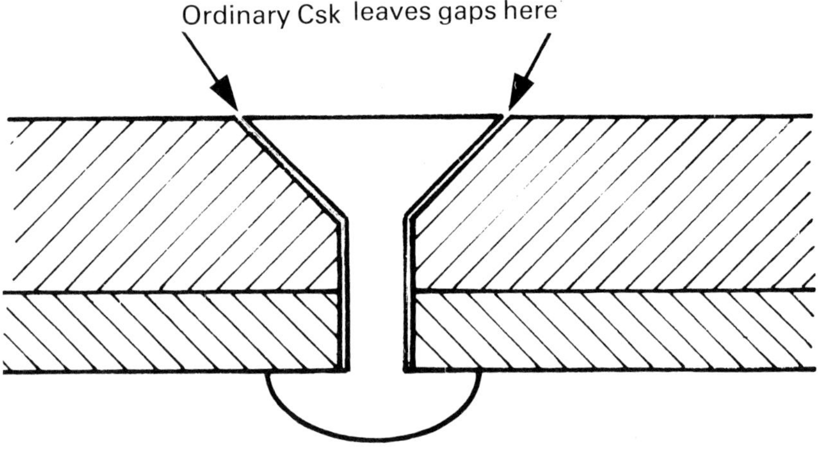

Ordinary Csk leaves gaps here

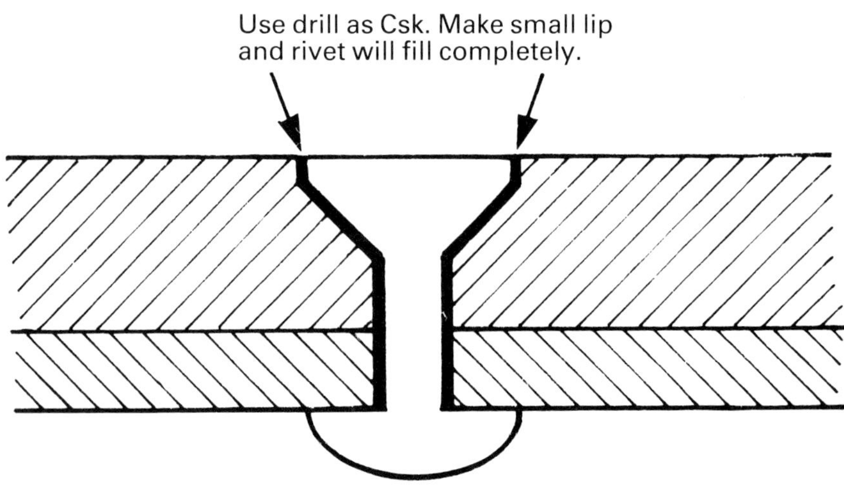

Use drill as Csk. Make small lip
and rivet will fill completely.

POP RIVETS

Pop rivets are popular in the motor trade. They can be used in model engineering, and are quick and easy to use. They consist of a hollow rivet with a pin through the centre. A special tool is used to grip the pin and pull it tight. This closes the rivet. When it is closed right up the pin snaps with a loud pop. The part that is left will fall out, leaving a well riveted job, if somewhat unsightly. The rivets are available for a range of sizes.

A pop rivet tool. Shown with a rivet in ready for use. Closing the arms together closes the rivet.

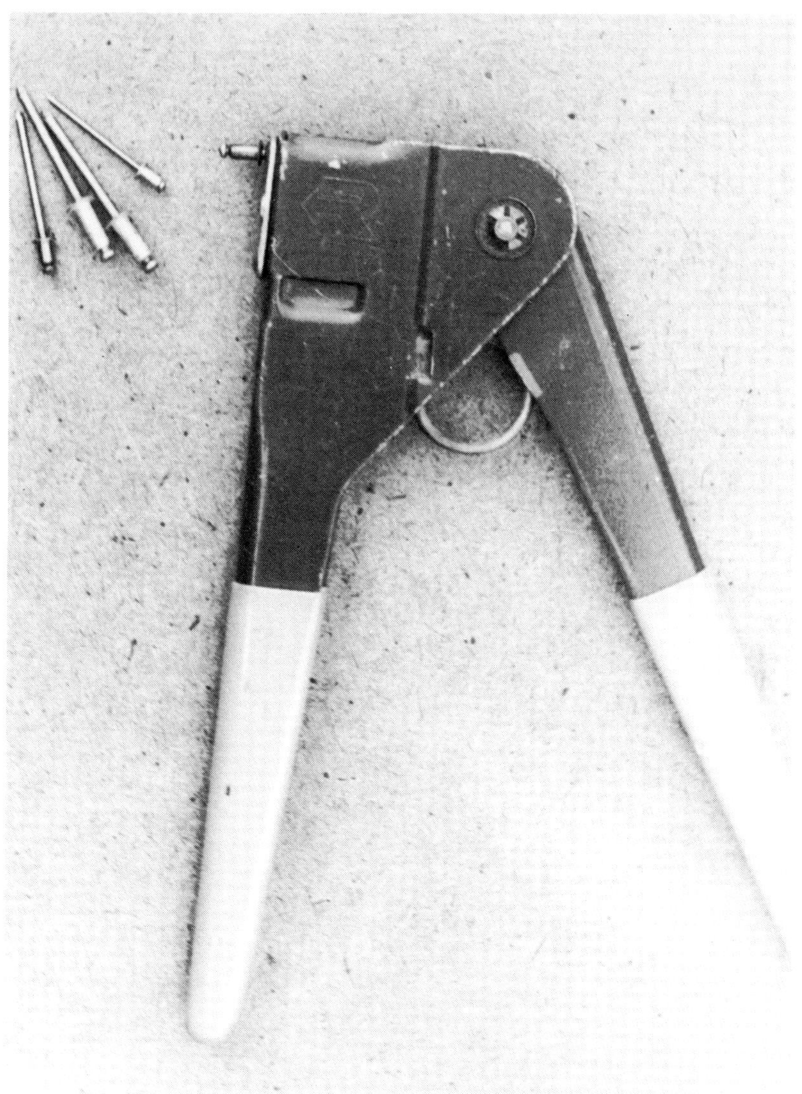

SIMULATING RIVETS

In small scale work as it is not always possible to get rivets of the correct size the metal is joined by other means and dummy rivets are put in for the sake of appearances. A simple tool is useful for this purpose and the sketch shows how to make it. Dimensions can be those suitable for your own purpose. The punch and die must be the correct size for the dummy rivet head that you require. Make a little allowance for the metal thickness when making the tool.

A simple tool for making dummy rivets.

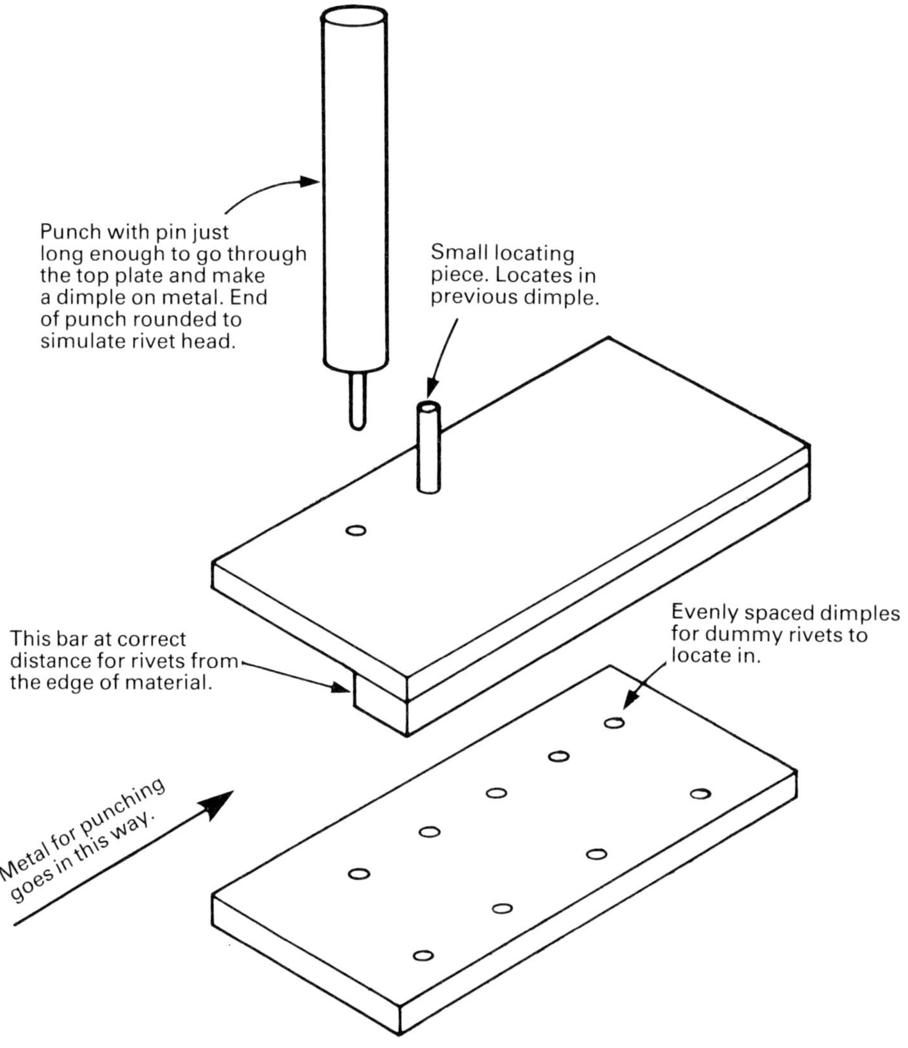

Punch with pin just long enough to go through the top plate and make a dimple on metal. End of punch rounded to simulate rivet head.
Small locating piece. Locates in previous dimple.
This bar at correct distance for rivets from the edge of material.
Evenly spaced dimples for dummy rivets to locate in.
Metal for punching goes in this way.
Bottom plate screwed to bar which gives room for material thickness.

PREPARING AND DRILLING

When preparing work for riveting the top piece of metal should be carefully marked out and all the holes drilled to the correct clearance size for the rivets to be used. Clamp the work that has been drilled to the piece to which it is to be riveted, and using the first piece of metal as a drilling jig, drill through the second piece at both ends, and one or two places in the centre. Rivet the two outside holes first, and then those in the centre that have been drilled. When you are certain that everything is alright, then drill the other holes and rivet through them, start from both edges and work towards the centre. Should any rivets be put in incorrectly, they must be drilled out and done again. To do this lightly centre punch the countersunk side in the middle, then drill through with progressively thicker drills until there is just the slightest rim of the rivet left. Then push the old rivet out with a punch. Do not try to drill full size in case the drill wanders and spoils the work. Where a rivet has a head on both sides you will need to file one of these off flush to the metal before punching and drilling. Trying to drill through a rivet head is a difficult proposition as they do not always lie true to the shank. Pop rivets that have been wrongly placed can be drilled straight through the hole that is left in them and the drill can be of the size required for riveting as almost invariably they lie true. They can also be removed by drilling with a larger size drill and stopping once the turned over section has been removed. More often than not they will then drop out.

16 *ADHESIVES AND FILLERS*

A few years ago you would not have found a chapter on adhesives in any book concerned with engineering. It is not so very long ago that the only sticky substances found in the workshop were a tar like compound that had to be lit at one end and the hot tarry substance dabbed on to the parts being stuck together. It was used for sticking items to face plates and that sort of thing and was never used for permanent work. But times change, and there are now no end of adhesives available, suitable for a whole range of purposes. There are numerous manufacturers of these substances and I could not hope to mention all the products by name. Most of them make similar ranges, and the adhesives are more or less based on the same formulae. I shall refer to the Loctite products in the text but I have no connection with Loctite other than using their products as a customer from time to time. This does not mean that I use only theirs, if another manufacturer's product is more conveniently available, I will use that. Loctite, however, do make a very wide range that makes description easy, and it is for that reason that I chose to use them as a guide to the many products available. There will from time to time be reference to other products where details of suitable adhesives are known.

TWO PART ADHESIVES

The first type of adhesives to be used in any quantity in industry were the epoxy based ones, which involved mixing the adhesive with a recommended quantity of special hardener. Such substances started life in earnest during World War 2, when they were used to good effect in the aircraft industry. As far as the model engineer is concerned they can be of considerable value if used properly. The first requirement of use, as with all adhesives, is clean, grease free parts to use them on. They will stick to glass, some plastics, wood, metal and fabric, and many other things as well. Once mixed they take some time to cure, and must be left alone during that period. They fill gaps very nicely and have great strength. I have regularly used them for insulating model locomotive wheels in gauge 'O' by cutting through all the spokes and filling the gaps so left with adhesive. Epoxy resins can be filed after setting. They cure better if warm and will withstand reasonably high temperatures. It is possible to purchase epoxy resins in a variety of curing times. The normal curing time is some 24 hours but there are types available that will cure within an hour or so. Basically it seems that the greater the curing time recommended by the manufacturer, the greater the strength of the bond. The best known of this type of adhesives are probably those manufactured by Araldite, in fact there is a tendency for an epoxy resin to be referred to as Araldite whether it was made by that company or not. A more recent development of the epoxy resin has been one with a liquid hardener which can make application considerably easier.

When using epoxy resin the work must be held whilst the resin cures in such a way that the parts will not move—the resins tend to liquefy before they set and will therefore move readily if not secured. However the parts must not be clamped together too tightly as they need a slight gap in order to get maximum strength. The recommended curing time can be speeded up, and the strength of the joint increased, by leaving the work to cure in a warm (not hot) oven or atmosphere. This is caused by the more rapid liquefying of the two parts.

Similar substances are sold as fillers, mainly for car body purposes, but they can have their uses in model making. They consist of epoxy resins and hardener plus an agent for filling, such as powdered metal with plaster. When hard they can be sawn, filed, drilled and tapped. As far as curing is concerned they behave in exactly the same way as the adhesives described above.

ONE PART ADHESIVES

These days the one part adhesives are becoming more and more popular for many situations. These work the opposite way to those just described, exclusion of air causing them to cure. They will not fill gaps. They are sold in a wide variety of strengths and specially made for different applications.

Screwlock (Loctite 222) is for lower strength holding of small screws and as such is ideal for the model engineer, for those places where a screw might shake loose when in use. It is easily loosened after use by applying pressure with a spanner or screwdriver. Nutlock (Loctite 242) is a similar product but somewhat stronger, probably too strong for nuts and bolts under 3 mm diameter. Studlock (Loctite 270) is stronger still and it is doubtful whether any model engineering application can be found for it. Where you might find it useful is if there are troublesome studs on any of your machinery that do not need to be undone except in absolute emergencies, then Studlock will hold them tight for you.

The next items in this range are known as retainers and whilst they have similar characteristics to the previously mentioned ones they have far greater strength. Loctite 601, simply known as Retainer, is very useful for model engineering purposes. I have used it with complete suc-

cess to hold the pieces of a crankshaft for a hot air engine together and for crankshafts on small marine engines. Slightly stronger still is Retainer High Strength, (Loctite 638) which as the name implies will hold parts together under tremendous load. I have seen a 7 ¼ in gauge model locomotive crankshaft held with it and it has been in operation for many years. If it is likely that parts may have to be taken apart at any time then Bearing Fit is the one to use, this is Loctite 641 and whilst retaining parts with terrific strength, a sharp blow with a small hammer will cause them to separate.

All these adhesives are intended for use in retaining smooth metal parts together. The work must be thoroughly cleaned before use and a special primer is sold to help the bonding. They must be used strictly according to manufacturers instructions, and the correct curing time allowed, if they are to work properly. Should there ever be a need to dismantle parts assembled with them then heating will allow this.

Opposite—*The crankshaft of this hot air engine is held together with Loctite retaining compound.*

Below—*On this marine engine sealing compound has been used on the pipe connections. Retaining compound was used for the crankshaft, and liquid gasket where the steam block meets the frame. To prevent them working loose in operation the pivot screws and big end bolts are held with low strength Nutlock.*

SEALERS

Pipe sealers and instant gaskets are useful for steam and water fittings. Pipe sealers come in two strengths. Loctite 577 Pipe Sealer should suit most model engineering purposes. Should there be any serious problems with sealing Hydraulic Seal (Loctite 542) is made to deal with hydraulic fittings up to a pressure of between 580 lb and 870 lb per square inch. Multigasket (Loctite 574) is a plastic like compound and can be used in place of the usual paper and composition gaskets. It is particularly useful where awkward shapes are involved. It is steam and oil proof as are the two sealing compounds.

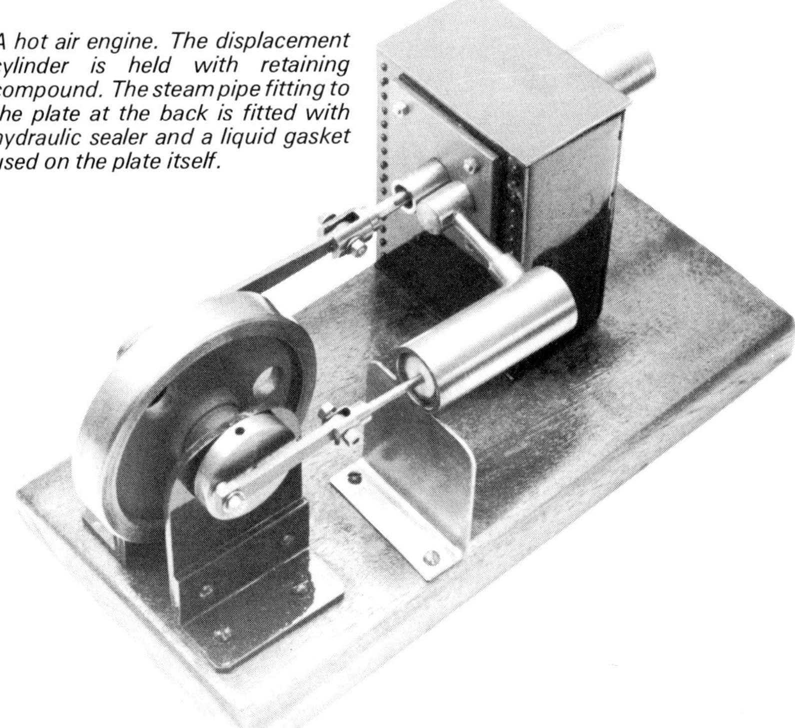

A hot air engine. The displacement cylinder is held with retaining compound. The steam pipe fitting to the plate at the back is fitted with hydraulic sealer and a liquid gasket used on the plate itself.

SUPERGLUES

Superglues, as they have come to be known, can be used for sticking non metallic surfaces to metal with considerable success. I do not think that they are ideal by any means as retaining compounds, although used properly they are very strong. They have one advantage over the adhesives mentioned so far in that they cure incredibly fast, so fast in fact that if you are not careful your fingers will finish up stuck firmly to the work. Most glue manufacturers make them and they are sold under a variety of names. I have seem them used for sticking plastic sheeting, which was used as window material, to metal superstructures and for sticking cloth and leather to metal, as well as for putting insulation material between strips of metal. For all these purposes they are very satisfactory.

PLASTIC METALS

At one time plastic metals were hardly worth any consideration, but so much advancement has been made in the field that they now can prove to be very useful indeed. Although I have no doubt that there are several firms making them Loctite and Devcon seem to be the main suppliers. They are usually available in a form of steel or aluminium, and can be used to build up parts that are undersize. Some are available as a form of putty and some in single part tubes. I have seen them used in industry to build up worn parts, but never for bearing surfaces, although the manufacturers could advise whether or not they are suitable for such. I have used a steel plastic metal to fill blow holes in a cast iron model locomotive cylinder, the cylinder having been virtually completed before the blow holes, which were quite deep, appeared. The locomotive has run extensively and there is no more sign of wear on the part that was filled than on the rest, despite the fact that the filler was applied to the bore and so it does take a considerable amount of wear. Most of these materials can be machined, on a lathe or a milling machine, and can be drilled and tapped as well as filed.

OTHER GLUES

Many glues are to be bought in multiple stores. All carry manufacturers instructions—read them before buying and make sure they will serve your purpose as in this type of shop there is little point in asking shop assistants or even the manager, they just buy what is suggested by the manufacturers. If in doubt try writing to the manufacturers, all of whom I have always found to be most helpful.

Before finally closing the subject there are a couple of other glues that might come in handy. The white wood glues are excellent when wood is to be stuck to wood. I also use them for making patterns for castings, with complete success. Impact adhesives too can be useful and if used properly they are very strong. They can be cleaned off with a solvent, and I find them useful for holding parts together to work on before final assembly. I use them for holding work to wooden blocks for awkward filing operations as well as holding items in place on face plates. The strength of impact adhesives is such that considerable force can be applied to work held together with them, without the parts separating or moving in any way.

This is a wooden wheel pattern for a 7¼ in gauge locomotive. It is made from pieces of wood laminated together with Evo Stick, white woodworker's glue. When dry, it has been turned, milled and filed.

17 *FINISHING*

To a degree the finishing of metal has been covered in the chapter on filing but there are some types of finish that have not been dealt with and as it is the finish which first creates the impression of quality on a model it is as well to think about the subject a little further. The use of very smooth Swiss files in good condition is without doubt the best way to impart a high quality finish to metal. The files soon become worn and scratchy, even if they are well looked after and this causes a loss of quality when finishing. Abrasive papers are probably the best alternative as far as flat surfaces are concerned. A coarse grade should be used first of all, then something a little finer, and finally the smoothest that you can get. The rubbing action should be done with the grain of the metal—this means lengthways on strip or bars, and usually lengthways on sheet. It does not matter too much in which direction you rub on the sheet, but care must be taken to keep all the abrasive marks in one direction. In the case of round bars the work should be held in a vice and emery cloth drawn in a semi circular motion round the bar. Rubbing lengthways on a round bar only creates a very dull finish.

Final work on metal can be carried out with an oil soaked strip of emery cloth. Great care must be taken to ensure the removal of all scratch marks. For an even finer finish, wire wool can be used. Again various grades are available and progressively finer ones should be used throughout the finishing. A little oil helps to achieve that extra little bit of scratch removal. The use of metal polish on steel is not worth considering in my opinion, but on brass and copper it will work wonders. It is also very good on aluminium, which tends to scratch badly with the coarser grades of emery cloth. The rule for that material is fine cloth, fine wire wool, and metal polish. Another substance that will impart a good finish on aluminium and all other metals is a mixture of household scouring powder and oil. Apply the mixture on a soft cloth and rub in one direction only. If the surface to be polished is wide then put a pad on a piece of wood, wrap a piece of cloth round it, and use the mixture in that way.

Sometimes you may have grooves or indentations in your project and whilst the main surface can be polished the groove remains rough and unsightly. Folding emery cloth and trying to rub that in the groove will only result in the edges becoming rounded, giving a worn appearance. The best way to deal with this situation is to get a piece of wood, any old bit will do, in fact I usually raid the firewood box for what I need. Cut the end roughly to the shape of the groove and, with the metal well oiled, rub the wood up and down in the groove, until it takes on the exact shape of it. Next apply some grinding paste, as sold for grinding valves in motor car engines. Rub this up and down the groove with the stick, and it will take the shape perfectly. Start with a coarse grade and finish with a finer one. Again, for that extra fine finish, use metal polish or the mixture of scouring powder and oil, applied with the stick.

Polishing mops can be purchased which revolve at high speed to polish the metal. Various grades of mop can be bought, and as usual it is as well to start with a coarse one and progress to finer ones. They are usually held on a tapered spindle and unscrew easily for quick changing. A polishing compound should be applied to the mop before using it, and these again are available in various grades as well as being made particu-

A polishing mop used in conjunction with a polishing compound will give a particularly smooth and bright finish to most metals. Goggles should always be worn when using a polishing mop.

larly to polish certain materials. Mops are highly efficient, probably too efficient—not only will they polish out all marks, but sometimes all detail as well. Even so if a mirror finish is required there is no better way of getting one. When teaching at school, I never ceased to be amazed at how long pupils were willing to spend mopping work, often to the detriment of the article being polished.

So far I have mentioned high polish finishes, but sometimes we may want something else. A piece of wooden dowl in a drilling machine with grinding paste, scouring powder, or something similar on the end, and rotated on the work in a series of circles gives an attractive finish, which I have heard called engine turning. If oil is incorporated in the abrasive medium it will also prevent rust to a degree. The use of scrapers can produce very attractive patterns on metal. Scrapers are just hard tools, which actually make fine scratches on the metal. This can be used to advantage. Scrapers are also used to get surfaces perfectly flat, and in this way two birds can be killed with one stone. A very nice grey finish can be obtained on steel, by using phosphoric acid, in conjunction with wire wool. Rub the acid hard in with the wool, preferably a very fine grade, and then before it dries rub the residue off with a piece of soft cloth. The result is a pleasant silver grey which will also prevent rust. Steel can also be blackened by heating it in an oil bath. This is somewhat messy for the home workshop, and if done in the garden is unlikely to endear you to the neighbours. It is possible to produce blackened brass and copper by chemical means but the processes are somewhat complicated.

Finally let us consider finishes that will not be seen. Sometimes we need two parts to mate accurately in such a way that they will be steam, water, or air tight without the use of gaskets. To do this both parts

The finish sometimes called engine turning leaves a pleasant pattern on the metal.

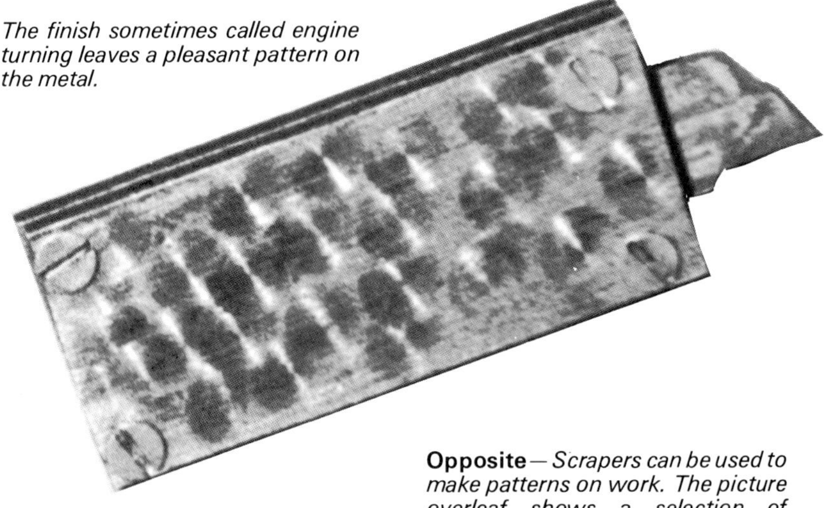

Opposite—*Scrapers can be used to make patterns on work. The picture overleaf shows a selection of patterns. A scraped finish can be very accurate.*

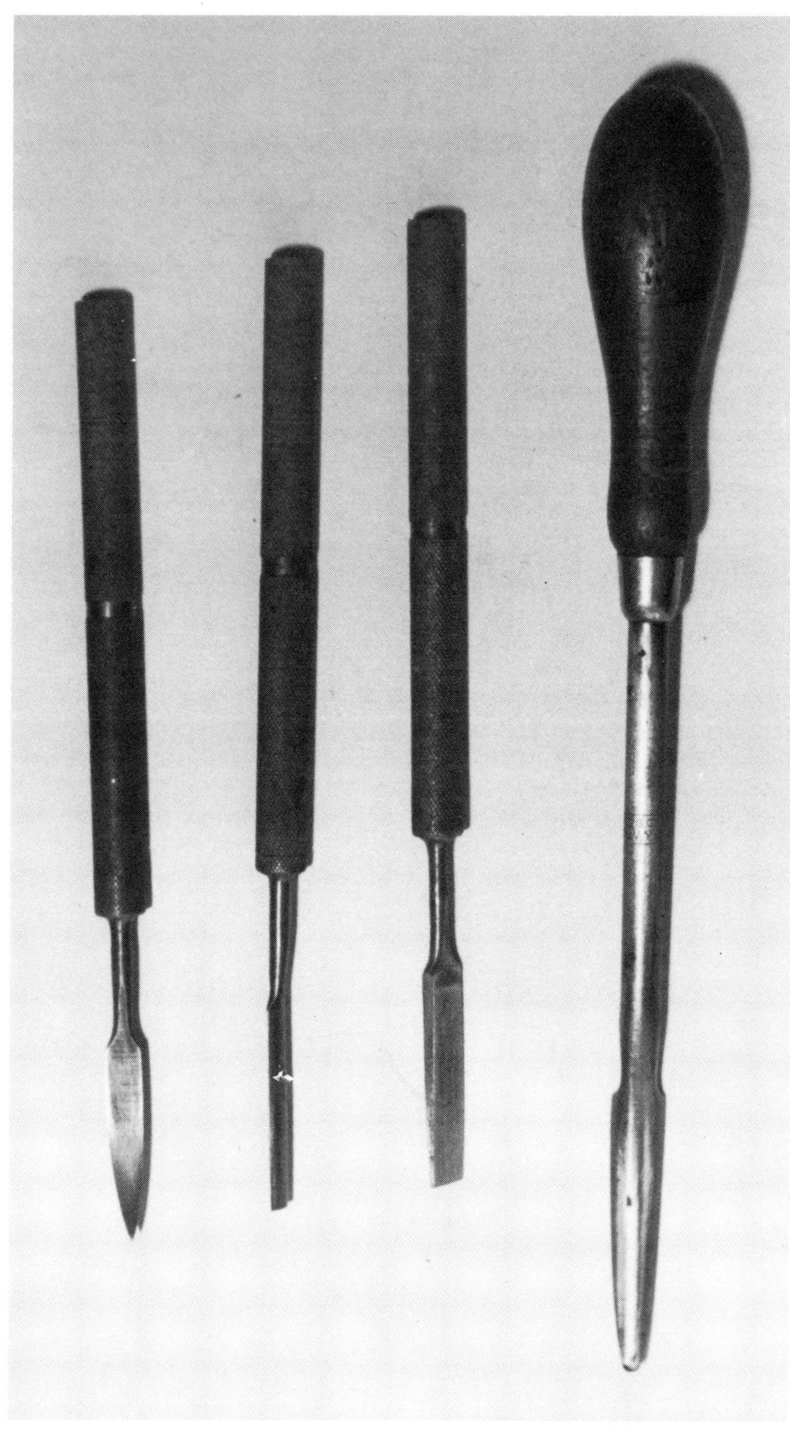

Two typical scraper patterns. One is made by moving the scraper in short straight pushes. Alternating the angle each time. The second by tristing the scraper as it is pushed. Care must be taken to keep the scraper absolutely flat or damage will result. It is probably as well to practice first on some scrap metal.

should be rubbed in a circular motion on emery cloth, whilst standing on a surface plate or piece of plate glass. Use very fine paper until all signs of unevenness disappear. Do not under any circumstances be tempted to rub them in a straight line or a rocking effect will be set up. When both parts are as near right as it is possible to get them they should be rubbed together, in position, with metal polish in between. This should ensure a perfect lapped joint. Again if possible rotate whilst the lapping is going on to prevent any rocking.

CONCLUSION

Working at the bench can be the most satisfying of pastimes. It can also be the most tedious. Assuming that the work is being carried out as a hobby, or indeed even if it is one's means of earning a living, it should be enjoyed. My experience is that it is only enjoyable if it is going well, and if that is the case we must make the work go well ourselves. Certainly if mistakes are made and work is slipshod then a feeling of virtual despair will creep in and there will be no pleasure in it. If the work is carried out properly then the finished article will be something to be proud of for ever.

To achieve this every operation must be carefully planned and thought out before starting. When reading a workshop drawing go over it several times, do not assume anything but check it all. Make sure the measurement you have read is the one that you want. It is easily possible to get the wrong one if care is not taken. Check that the part of the drawing you are looking at marries correctly with the part to which it will be attached. Drawings often contain mistakes, so try and check that the one you are referring to does not before taking for granted what is written on it.

When the measurements are to be transferred to the work check two or three times that you have measured correctly. Make sure that scribers held in scribing blocks are firmly secured, and go lightly with the hammer on the centre punch, you may yet need to change the position of the centre punch mark. Do not try to hurry or take short cuts, and remember to use only two datums.

If filing or sawing, take it slowly. Hack saws have a remarkable ability to cut curved lines, although in theory they are not capable of doing so. Old and worn blades are more likely to do this than are nice sharp new ones. Use the correct blade for the material. Keep your files clean and do not let them rub against each other. Remember to keep different ones for different grades. Both hacksawing and filing operations should be carried out slowly, hurrying inevitably means disaster.

Keep drills sharp and use them at the correct speed. Take it easy when tapping and in both operations do not forget to use a cutting fluid. Also in both cases be gentle—taps and drills are expensive but what is probably more important is the fact that a broken tap or drill is a disaster.

When finishing metal carry out the work and then put it down for an hour or two. When it is picked up again it is almost certain that the finish will not be good enough and will need a little more work. By doing this the article will be much better finished in the end than if the original work is taken as being suitable. Finish is important, it is the thing that sets the work off and gives the all important first impression.

If all this sounds like rather a lot to think about do not worry as it all boils down to the fact that whatever you are going to do in the way of metalwork, if care is taken the result will be good and you will find it most satisfying and enjoyable.

APPENDIX *SAFETY*

Although little machinery is used in general benchwork operations there is still a need to have regard to safety precautions in the workshop and every possible care should be taken in this respect. It should start with your personal dress. At no time should sandals with open toes or soft topped footwear be worn whilst working, there is always a danger of heavy items, possibly hot ones, falling down and if protective footwear is not used nasty injuries can occur. Likewise suitable outer clothing, in the form of an overall, should be worn when there is any danger of swarf being generated. Hot swarf landing on bare arms or neck can be a very painful experience indeed.

When welding or brazing make sure that suitable goggles or a face shield is used. If there are sparks coming from the work wear some form of headgear—hair that has a dressing on it can catch fire as such substances often contain grease or other inflammable chemicals. If there is danger of fumes from any form of work, and this particularly applies to welding and brazing, then wear a mask which will offer some protection. Do not remain long in the atmosphere containing the fumes, but go into the fresh air at frequent intervals. Cutting oils will often give off fumes that can cause considerable irritation or even some form of skin disease so take care not to inhale these fumes. Some plastic materials can also give off toxic fumes and again care should be taken, some can be particularly offensive and dangerous when heated. Do not therefore smoke either a pipe or cigarette when working with them, a small piece of the material caught by the lighted pipe could give off enough fumes to make you feel quite ill. Smoking in a workshop should be discouraged anyway, many cutting fluids are highly inflammable and hot ash can cause them to burn. Some means of extinguishing a fire should always be kept in the workshop and an escape route should be available in case of mishaps.

Be careful with sharp tools which can cause nasty cuts. If you do cut yourself clean and dress the wound at once, remember the tool contains not only a sharp edge but a fair amount of dirt as well. Metal splinters in the hands can be a problem and it seems that no matter what you do you will somehow or another get splinters. They should be removed as soon as possible. Whilst it is difficult to avoid getting them they occur mostly when cleaning up so wear an old pair of gloves for this purpose if possible. Avoid wiping tools and benches down with the hands. Use a piece of cloth, preferably something that can be thrown away afterwards. If it is possible to have a vacuum cleaner in the workshop this can save a lot of wear and tear on your hands.

Old and badly worn tools are a hazard. Do not use files with broken handles, or worse still without handles at all. Do not use punches that have burred over at the top as there is a good chance of the hammer slipping off them, grind the top smooth if required. Do the same with hammer heads—burred over hammer heads will sometimes have chips of metal fly off them or a piece will break off and the hammer will slip. Grind away any metal that is burred over before use. If using a drilling machine then use the guard if fitted. Keep loose sleeves away from all drills which are in operation. If using slotted screws and the slots are worn throw them away and get new ones. They are a source of danger and anyway will be difficult to remove in the future. Keep any chemicals that may be in the workshop in sealed and well labelled jars. Remember if mixing acid and water that, whilst acid can be added to water, to try and add water to acid could result in a small explosion with the acid being thrown up in the air.

Finally try and fix up some sort of emergency warning system in the workshop, so that in the event of an accident help can be summoned quickly.

Decimal equivalents of fractions of an inch

Fraction in	Decimal in	Fraction in	Decimal in	Fraction in	Decimal in
1/64	.0156	11/32	.3437	43/64	.6718
1/32	.0312	23/64	.3593	11/16	.6875
3/64	.0468	3/8	.3750	45/64	.7031
1/16	.0625	25/64	.3906	23/32	.7187
5/64	.0781	13/32	.4062	47/64	.7343
3/32	.0937	27/64	.4218	3/4	.7500
7/64	.1093	7/16	.4375	49/64	.7656
1/8	.1250	29/64	.4531	25/32	.7812
9/64	.1406	15/32	.4687	51/64	.7968
5/32	.1562	31/64	.4843	13/16	.8125
11/64	.1718	1/2	.5000	53/64	.8281
3/16	.1875	33/64	.5156	27/32	.8437
13/64	.2031	17/32	.5312	55/64	.8593
7/32	.2187	35/64	.5468	7/8	.8750
15/64	.2343	9/16	.5625	57/64	.8906
1/4	.2500	37/64	.5781	29/32	.9062
17/64	.2656	19/32	.5937	59/64	.9218
9/32	.2812	39/64	.6093	15/16	.9375
19/64	.2968	5/8	.6250	61/64	.9531
5/16	.3125	41/64	.6406	31/32	.9687
21/64	.3281	21/32	.6562	63/64	.9843

Conversion — millimetres to inches

mm	in	mm	in	mm	in	mm	in
1	.039	26	1.024	51	2.008	76	2.992
2	.079	27	1.063	52	2.047	77	3.031
3	.118	28	1.102	53	2.087	78	3.071
4	.157	29	1.142	54	2.126	79	3.110
5	.197	30	1.181	55	2.165	80	3.150
6	.236	31	1.220	56	2.205	81	3.189
7	.276	32	1.260	57	2.244	82	3.228
8	.315	33	1.299	58	2.283	83	3.268
9	.354	34	1.339	59	2.323	84	3.307
10	.394	35	1.378	60	2.362	85	3.346
11	.433	36	1.417	61	2.402	86	3.386
12	.472	37	1.457	62	2.441	87	3.425
13	.512	38	1.496	63	2.480	88	3.465
14	.551	39	1.535	64	2.520	89	3.504
15	.591	40	1.575	65	2.559	90	3.543
16	.630	41	1.614	66	2.598	91	3.583
17	.669	42	1.654	67	2.638	92	3.622
18	.709	43	1.693	68	2.677	93	3.661
19	.748	44	1.732	69	2.717	94	3.701
20	.787	45	1.772	70	2.756	95	3.740
21	.827	46	1.811	71	2.795	96	3.780
22	.866	47	1.850	72	2.835	97	3.812
23	.906	48	1.890	73	2.874	98	3.855
24	.945	49	1.929	74	2.913	99	3.888
25	.984	50	1.969	75	2.953	100	3.937

INDEX

Other PSL books for model engineers

Introducing the Lathe

Stan Bray

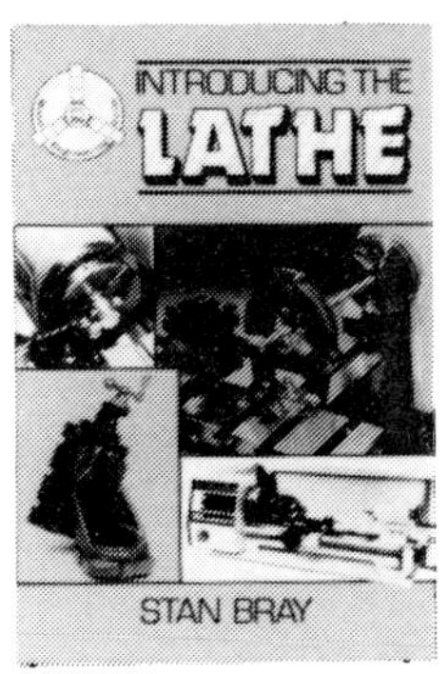

The lathe is probably the most versatile tool in any home workshop, and here Stan Bray not only explains how to choose one to fit your needs and your pocket but also how to use one safely. Cutting tools, holding the work, tool posts, drilling and boring, threading, milling and many subsidiary operations are also covered. This book is the essential complement to *Introducing Benchwork.*

96 pages, illustrated, paperback.

Modelling Stirling and Hot Air Engines

James G. Rizzo

This title not only describes the components of a typical hot air engine but also shows how model engines of varying complexity can be constructed from essentially scrap materials. The theory and evolution of hot air engines from the days of the Reverend Stirling to the present is also described. These delightful little machines make excellent projects on which the budding model engineer can 'cut his teeth'.

192 pages, illustrated, paperback.

Modelling Ships in Bottles

Jack Needham

You don't even need a workshop to construct beautiful model ships in bottles which will amaze your friends. Learn all the secrets of a master modeller from one book! Step by step diagrams and photographs make every stage of construction easy to follow, and the book also shows how objects other than ships can be modelled in bottles as well.

168 pages, illustrated, hardback.